Funding Transport Systems

Related Pergamon books

CULLINANE & STOKES	Issues in Rural Transport
DAGANZO	Fundamentals of Transportation and Traffic Operations
ETTEMA & TIMMERMANS	Activity-Based Approaches to Travel Analysis
HENSHER, KING & OUM	World Transport Research: Proceedings of the 7th World Conference on Transport Research (4 volumes)
LESORT	Transportation and Traffic Theory
ROTHENGATTER & CARBONELL	Traffic and Transport Psychology: Theory and Application
STOPHER & LEE-GOSSELIN	Understanding Travel Behaviour in an Era of Change
TURRO	Going Trans-European

Related Pergamon journal

Transport Policy
Editor: P. B. Goodwin

Free specimen copies available on request.

Funding Transport Systems

A Comparison Among Developed Countries

By

D. NAKAGAWA and R. MATSUNAKA
Department of Civil Engineering Systems
Faculty of Engineering
Kyoto University

Pergamon

Amsterdam   New York   Oxford   Singapore   Tokyo

U.K.   Elsevier Science Ltd, The Boulevard, Langford Lane, Kidlington, Oxford
            OX5 1GB, U.K.

U.S.A.   Elsevier Science Inc., 655 Avenue of the Americas, New York 10010,
            U.S.A.

JAPAN   Elsevier Science Japan, Higashi Azabu 1-chome Building 4F, 1-9-15,
            Higashi Azabu, Minato-ku, Tokyo 106, Japan

First edition 1997

Library of Congress Cataloging in Publication Data
A catalog record for this book is available from the Library of
Congress

British Library Cataloguing in Publication Data
A catalogue record for this book is available from the British Library

ISBN 0 08 043071 6

Typeset, Printed and Bound by The Charlesworth Group, Huddersfield, UK

CONTENTS

ILLUSTRATIONS AND TABLES

ILLUSTRATIONS

TABLES

Chapter 6

Chapter 1

Introduction

Transportation systems play a central role in supporting both economic activity and social life, but despite major capital investment in the past most countries are finding that their transport systems can barely keep up with demand. Unless transport is to become a constraint on further economic development, ways must be found to improve them. A key aspect of this problem is funding.

Methods of financing range from the orthodox, such as generating tax revenues and collecting fees from users, to the more esoteric, such as placing the burden on specific beneficiaries. From the viewpoint of the burden theory of financial resources, the question of who should pay is of deep concern and is related to the most basic questions of how to assess the role and purpose of transportation improvements. If it is believed that improvements will only result in increased user convenience, then users should bear the cost, however, transportation systems are part of the social infrastructure for all industries and activities, as well as having a social welfare dimension in ensuring freedom of movement. In addition there may be external effects such as capital gains accruing to local land owners. Finally, the question of financing is complicated by such issues as the external costs of environmental impact and the need to distribute the burden across generations.

The aim of this book is to discuss basic concepts and practice of financing of transportation systems. In the first half, after describing the theoretical basis of burden, the policies and financial systems of some developed countries are introduced and compared analytically. In the second half, a methodology for comparing the structure of financial resources is developed, and actual investment per group of contributors is calculated.

In the first half of the book, the focus is on each country's position on the following issues: 1) the responsibility of the public sector and the role of the private sector; 2) the balance of burden covered by general funds and by users; and 3) The balance of burden covered by present financial resources and by debt. The second half shows how such national policies are reflected in their financial resources. Here, after a detailed cross-country review of transportation finances, a methodology is developed for an international comparison of financial resources for transportation system improvement. This methodology utilizes a concept of "actual contributors" as a way of cross-country standarization so avoiding superficial comparisons based on incompatible statistics.

Previous studies used partial and fragmentary investment data but this book includes comprehensive data covering all roads, railways, and airports. When making international comparisons, an attempt has been made to apply unified standards rather than comparing superficial statistical values, since methods of recording and calculating statistics vary from country to country. Much work remains to be done in this area if discussion on appropriate financing of, and optimum improvements to transportation is to progress.

Based on the purposes mentioned above, the contents of this book are as follows.

In Chapter 2 the theoretical basis of burden is described, the basic concepts and important financial systems required for improving transportation systems are reviewed, and some recent characteristic systems are introduced. In Chapter 3 the transportation policies and financial systems for the roads, railways and airports of five developed countries (Germany, France, United Kingdom, U.S.A. and Japan) are analysed in relation to financial resources, then in Chapter 4 a method for an international comparison of financial resources is introduced. This incorporates the concept of "actual contributors" and leads to a classification of the financial resources needed to improve the transportation systems in each country.

The amounts of investment and its distribution between roads, railways and airports are calculated for five developed countries in Chapter 5, then in Chapter 6 the actual shares are calculated for each group of contributors. These groups include "national" and "local tax-payers" who contribute to general funds; "users" who pay fees, including tolls, fares and users' taxes; "future users" who contribute through the burden of debt repayment, and "other specific beneficiaries," who pay special taxes or contributions. These results are used in Chapter 7 to compare actual financial resources among the developed countries by looking at the degree of dependence on debt, the ratio of special-purpose taxes imposed on the users, the ratio of national and local taxes to the total investment amount, and so on.

Based on this analysis Chapter 8 considers prospects for future improvement of transportation systems.

CHAPTER 2
BASIC THEORY AND POLICIES

2.1 THEORY OF FUNDING

Transport can be funded from a variety of sources ranging from taxation to user charges, but ideally a combination of methods should be used. In addition:

1) investment should match need;

2) projects should be efficiently realized;

3) the burden should be fairly and equitably distributed.

The use of general public funds derived from taxation may be effective in achieving a fair distribution of burden, but operational efficiency requires a market mechanism driven by user choice. The type of funding chosen will depend on the nature of a country's transport system and its level of economic development.

Historically, transport infrastructure has been substantially funded by the public sector but there is now an increasing tendency to rely on the private sector. Deteriorating national finances make it difficult to find sufficient funds for transport development but, at the same time, environmental considerations are persuading governments of the need to promote modal shifts. All these issues contribute to the complexity of the funding problem. The first part of this chapter describes basic theories relevant to the discussion of these problems.

2.1.1 The theory of welfare economics

Theoretically the market mechanism should ensure improvements in transportation for all types of goods and services, however, the market alone may not be sufficient to realize the three conditions of investment described above, i.e. adequacy, efficiency, and fairness. Market failure occurs because of the presence of several constraints of which the most important are external diseconomies and economies, and fundamental social rights.

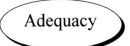

Adequacy

The systems must be capable of realizing an adequate scale of investment

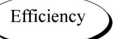

Efficiency

The systems must allow efficient realization of the project

Fairness

The systems must achieve a fair and equitable burden

Figure 2.1. Requirements for funding systems.

(1) External diseconomies (outside the transportation market)

This term refers to those costs arising from transportation but not necessarily paid for within the transport market. One of the most significant examples of an external diseconomy is environmental damage. The adverse effects of the use of automobiles, such as air pollution and noise, represent a heavy social cost that is not necessarily borne by individual car users. According to economic welfare theory, excessive traffic volume is the result of a difference between actual user cost and total social cost. To maximize welfare the two costs should be equalized. There are several ways of doing this, e.g. by applying an "environment tax," road pricing (as practiced in Oslo and Singapore), fuel tax. A less direct way of curbing automobile traffic is by subsidizing alternative modes of transport such as the railway or buses.

(2) External diseconomies (within the framework of the transportation market)

The term external diseconomies is sometimes used in a narrower sense to mean the costs that users impose upon each other. To give an example, car drivers typically perceive the time they spend at the wheel as a cost, but they fail to consider the cost in time that they impose on other car drivers as a result of their entry into the traffic flow. In other words, the marginal time cost perceived by individual users is less than the sum of the marginal cost of all users (ref. Expression in Figure 2.2), and this failure in perception gives rise to excessive traffic volume.

One method of transferring the costs to users is by means of a "congestion tax". Road pricing can be seen as a way of dealing with external diseconomies both inside and outside the framework of the transportation market: it can be perceived as an environmental protection measure or as a measure for reducing road congestion.

(3) External economies

External economies refer to benefits accruing outside the transport market, for example, land values might increase when access is improved. Sometimes these beneficiaries can be identified but in many cases this is more difficult. If they are clearly identified it is possible to recover the cost by way of direct taxes or charges. Where identification of individual beneficiaries is difficult, as in the case of improvements to the environment which benefit the population at large, the cost may be met from general public funds. A property tax used to fund improvements to transportation is one method.

(4) Fundamental social rights

Fundamental social rights are rights consensually guaranteed by society. Such rights might include the right of access for everyone including the aged, the handicapped and those living in remote areas. A pure market system would not

> **External Diseconomies**
> (Outside of the transportation market)

Social Marginal Cost
> Personal Marginal Cost

Method of Distributing Burden

- Direct Method
    Added charges : environment tax,
            road pricing.

- Indirect Method
    Reduce demand on the demand curve :
        Subsidies to such alternative
        modes as railway or bus.

Cost

Social MC

Personal MC

Dead Weight Loss

Appropriate   Personal   Demand
Demand   Demand

> **External Diseconomies**
> (Within the framework of the
> transportation market)

Method of Distributing Burden

- Added charges : congestion tax,
            road pricing.

Total Travel Cost $= T(x) \cdot x$

$$\text{Marginal Cost} = \frac{\partial}{\partial x}\big(T(x) \cdot x\big) = T(x) + \frac{\partial T(x)}{\partial x} \cdot x$$

Personal   **External**
Cost   **Diseconomies**

x : Travel demand
$T(x)$: Traveling time

> **External Economies**

Method of Distributing Burden

- Return of development profits
    (Beneficiaries' shares)

ex. Increase of the real estate value

Figure 2.2. Externalities.

provide these services so there is the need for some kind of public support (e.g., subsidization).

2.1.2 Fundamental theoretical issues

(1) Pricing and funding: two theories
Welfare economics treats externalities within the framework of pricing theory, i.e. it is used to determine the set of charges which maximizes social surplus. Optimum charges are found at the point where the marginal social cost curve crosses the demand curve. "Congestion tax" and "Two-part tariff" are examples of this approach.

The theory of funding is used to make decisions on the allocation of resources between, for example, road construction or improvements to public transport. Marginal cost theory can handle fixed costs by determining a long-term marginal cost curve, although in practice it is difficult to establish the crossing point between the long-term marginal cost and demand curves.

(2) Benefit measurement and funding theory
The theory of funding can be applied relatively easily when the size and incidence of benefit is known. In this case all that is needed is to match burdens to beneficiaries. In reality, however, things are rarely this simple. For example, a transportation system initially benefits its users, but benefit is gradually extended to real estate owners in the form of property value appreciation. The theory of funding provides an approach to equalizing benefit and burden taking account of this process.

(3) Efficiency and funding theory
In recent years a number of countries have privatized their major railways and institutionalized the construction of private toll roads in the expectation that higher efficiency will be found in the private sector than in the public sector. Improved efficiency results from competition and the operation of the profit-motive. Privatization of public transport can be justified if increased efficiency resulting from the market mechanism can be reconciled with the issues of externality, fundamental social rights, equity and inter-generational fairness. The relationship between privatization and infrastructural improvement is discussed later in Chapter 3.

(4) Long-term benefits
Transportation projects provide long-term benefits across several genera-tions thus justifying long-term debt. Characteristically this is achieved by issuing bonds. Reimbursement may come from general public funds or by means of an

incremental property tax on real estate appreciation (see TIF, Chapter 2, Section 4).

2.1.3 Characteristics of main funding sources

Main funds for improving transportation comprise public funds derived from general taxation, funds from user charges, and funds from debt. Each source has its advantages and disadvantages which can be discussed within the framework described above:

(1) Public funds: merits and demerits

The use of public funds is the preferred means of financing transportation when externalities are present or when fundamental social rights need to be guaranteed. Use of public funds is a political decision and can satisfy the principle of fairness, unless the decision-making system is inadequate and arbitrary. Generally speaking, work that is state funded involves little incentive for profit-making and for this reason tends to be relatively inefficient.

(2) User charges: merits and demerits

The size of revenues from user charges depend on customer choice so pricing must be competitive. By the same token, unnecessary investment and waste of resources are discouraged. But if a transportation system depends entirely on this revenue, profitability becomes the sole basis of decision-making and social benefits and costs are ignored. External diseconomies lead to excessive investment, and external economies result in insufficient investment. Such investment is likely to be concentrated in large cities where profitability is higher than in less densely populated areas.

(3) Long-term debt: merits and demerits

Although the distribution of burden between present and future generations may achieve a degree of fairness, future generations cannot be consulted. The level of investment is based on uncertain cost and benefit estimates and the accuracy of these estimates affects the quality of decision-making. Excessive investment will be undertaken if future benefits are overestimated; inadequate investment will result from underestimation. Uncertainty makes effective decision-making a highly difficult task.

To sum up, the best approach from a theoretical point of view is to use funds derived from the general account if externalities exist, debt if improvements benefit future generations, and the remaining costs should be born by users. In reality, however, it is far from easy to determine the ideal combination of funds because it is no simple matter to estimate those factors ignored by the market or those involving some degree of uncertainty. How this particular problem is perceived and dealt with by various countries will be studied in some detail in the following section.

2.2 CONCEPT FOR FINANCIAL RESOURCES

2.2.1 Transportation authorities

Different countries have different institutional arrangements for managing their transport systems. Characteristically there will be a combination of public and private provision but the balance, which largely depends on existing facilities, can vary greatly. Construction tends to be seen as a state responsibility, especially in Germany and France.

In Germany the "Basic Law (Grundgesetz, Article 87)" states that major transportation byways (roads, railways and waterways) belong to the federal treasury so the federal government has responsibility for improvements. Although the German Railways Corporation (Deutsche Bahn Aktien Gesellshaft: DBAG) was established in 1994 as a private operating business, construction and investment remain under federal control. International airports are being improved and operated by a limited number of private companies.

The French Basic Law for Domestic Transportation (Loi d'Orientation des Transport Intérieurs: LOTI), gives state organizations responsibility for construction, maintenance and management of fundamental transportation facilities, but construction and maintenance of toll expressways is entrusted to mixed-economy or private companies. These organizations were established by investment from the public sector so they are not entirely private, but their organizational structures would classify them as belonging to the private sector.

In the United States and the United Kingdom the private sector is being considered for a role in toll road construction, but, to date, France is the only country to have used this system extensively. With few exceptions, railways in Japan belong to the private sector which has responsibility for both construction and operation.

The operational side of the national railways of Japan, Germany, and the United Kingdom is increasingly being transferred to the private sector, though the construction of basic facilities remain in the public sector. This is known as the separation of infrastructure and operations. The construction and maintenance of basic transport facilities are seen as part of the social infrastructure and are therefore state responsibility, whereas private enterprise is best placed to handle operations.

The Swedish National Railway (Statens Jarnvagar) was the first to execute this idea in its entirety. Germany and Britain have followed suit though the details of their systems vary, whereas the Japan Railway companies (JR) maintain responsibility for both construction and operations.

2.2.2 Financial resources

Even when the state is responsible for investment, money from the general fund may be supplemented by revenue from users in the form of tolls, fares and airport facility rental charges, as well as taxes on automobile and aircraft fuels. All countries use some combination of these approaches depending on the basic concepts underlying transportation.

These concepts are: 1) the balance between market competition and policy intervention; 2) recognition of the range of benefits of transportation improvement; and 3) determination of the primary purpose of transportation improvement.

The first two points are clearly linked. In recent years it has been emphasized that competition is fundamental to transportation improvement and the increasing privatization of transportation operations is a result of this shift in emphasis. However, if the benefits from improvement are subject to externalities, social compensation becomes necessary. Policy intervention is required to promote social welfare, energy conservation, and environmental protection. In addition, an argument can be made for customer choice of transport mode.

The political decision on free market versus intervention will depend on the nature of the benefits from the improved system. There are often stakeholders who are not compensated or charged by the market, such as those who have suffered adverse environmental impact, or others who have benefited from a rise in property value.

The choice of financial resources also depends on the incidence of these costs and benefits. If users are the main beneficiaries then usage fees are chosen. If most benefits accrue to non-users then, where the beneficiaries can be identified, they should shoulder the burden. Otherwise general funds are chosen. In practice it is difficult, both to determine where the benefits and costs fall, and how to calculate them.

Consideration of the balance between competitive principle and policy intervention.

Recognition of the range of the benefit of transportation improvement.

Determination of the primary purpose of transportation improvement.

Figure 2.3. The points forming the basic concept.

The third point relates to the purpose for which these improvements were carried out. For example, if transportation is improved to correct domestic regional differentials or to reinforce international competitiveness, i.e. for strategic reasons, these goals will not be fully realized by the market mechanism, but will require public intervention.

In Germany it is assumed that the fundamental transportation system of road, rail and waterways should be improved as a part of the social infrastructure thus placing the burden on the federal government. However, a considerable portion of state funding comes from car-related taxes.

In France, basic transportation policies covering all railways, roads, inland waterway, and aviation are specified in the "Basic Law for Domestic Transportation (Loi d'Orientation des Transport Intérieurs: LOTI, 1982)" as are individuals' "transportation rights (droit au transport)." The largest share of the burden of both roads and rail is born by general funds. Because the ratio of toll roads is high, the burden on users is comparatively large. Another element of the French system is the "Versement de Transport," which links costs and benefits of non-users. Well located offices are taxed regardless of transportation usage. The important point of the "Versement de Transport" is that it is a specific fund for improving transportation. (The "Versement de Transport" is explained in 2.3.2.)

By contrast, Japan's 11th Report of the Transportation Policy Council declared that transport improvement costs should be born by users. This is especially true of the railways. Apart from subway lines, most of the railways in urban areas are improved and operated by the private sector. The assumption that a railway is public property is not clearly defined in the financial system. Roads, such as expressways, are also funded by users; expenditure for both construction and operation come from self-supporting systems where expenditure is covered entirely by usage fees. This system is generally applied to expressways, railways and airports.

Compared to the countries mentioned above, the United States and the United Kingdom have adopted intermediate strategies. Traditionally the public sector has taken responsibility for transportation, however, the idea that the private sector should take over this role has become popular in recent years as evidenced by the construction of toll roads.

All countries fund their transport systems through a mixture of public funds and user revenue. In general, efficiency is expected to rise when users pay, i.e. through the market mechanism, and fairness is enhanced through the use of public funds. Therefore the balance of financial resources is different in each country depending on political attitudes towards externalities.

2.3 BASIC FUNDING SYSTEMS

This section looks at the various ways in which different countries fund improvements to their highways, railways and airports through car user taxes, tolls, public subsidies and private finances.

2.3.1 Automobile users' taxes

All countries collect taxes for the possession and use of automobiles, but they differ in their methods of treating this revenue: it may be used solely for funding the highway system or for funding transport more generally: it may even be added to the Exchequer's general funds and used in ways quite unrelated to transport.

In Germany legislation for automobile users' taxation was introduced in the Federal Long-Range Road Act (Bundesfernstraßengesets) (1953), Transport Finance Act (Verkehrsfinanzgesets) (1955) Road Construction Finance Act (1960). Since 1973 these monies have been available for both road and non-road transport development.

In the United States the Highway Trust Fund was established by a road revenue act of 1956 to promote improvement of the Interstate and Defense Highway; money comes from automobile users' taxes, subsidies from the general account and from toll road fees. Many states are now establishing their own highway trust funds.

In Japan specific roadway improvement funds were introduced in 1958 on the basis that beneficiaries and those who cause wear and tear should foot the bill. Thus most funding comes from automobile users' taxes, and very little from general funds.

There are also countries that treat automobile users' tax revenues as a contribution to general funds. In France a specific fund for roadway investment (Funds Spécial d'Investissement Routier), financed by a fuel tax, was established in 1951 but since 1981 it has been merged into general funds. Similarly in the United Kingdom, a road development and improvement act of 1909 placed road

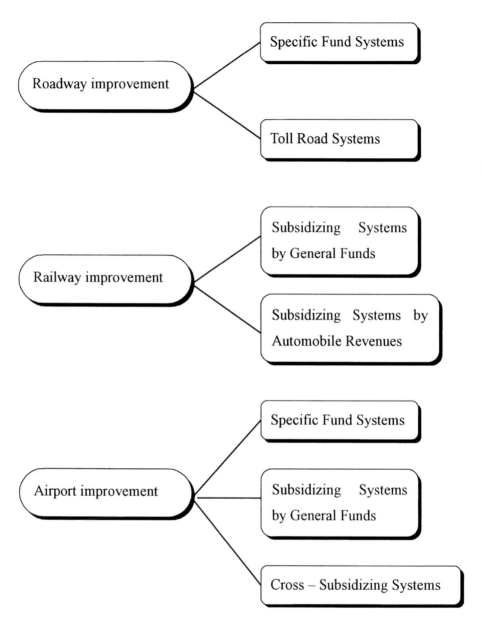

Figure 2.4. Important systems.

Table 2.1. Specific fund for roadway improvements.

	Germany	France	U.K.	U.S.A.	Japan
Usage of Automobile Taxes	Specific Fund (50%)	General Fund	General Fund	Specific Fund	Specific Fund
Establishment of Specific Fund	1955	FSIR 1951 (Abolished in 1981) FSGR 1982 (Abolished in 1986)	1909 (Abolished in 1937)	1956	1956

and car taxes in specific funds for roadway improvement until 1937 when the system was discontinued.

Automobile users' taxes may be paid into general funds and investment in roadways subsidized from general funds, so the existence or otherwise of specific roadway improvement funds is not substantially related to the size of investment in road construction. However, where such funds do exist financial resources will be available independently of national financial budgetary conditions, with the advantage that investment can be made without the need of political negotiation and compromise. As a consequence heaviest investment tends to occur in countries which have specific funds for roadway improvement.

2.3.2 Toll road systems

Internationally the use of toll roads is limited, but recently attitudes have been changing and it is likely that their number will increase in the future.

In France most of the expressways are toll roads. Prior to their introduction France's major road system was undeveloped in comparison to other European countries, but this is no longer the case. Construction and maintenance of toll roads is carried out by a group of semi-governmental corporations, "Sociéte d'Economie Mixte (SEM)" or by private companies funded by banks and public bodies. Money for toll expressways is prepaid by the French Expressway Organization (Autoroute de France: ADF), local governments and bonds issued by companies, and is reimbursed from toll revenues.

Table 2.2. Toll road systems on expressway.

Germany	France	U.K.	U.S.A.	Japan
A fee has been charged to only heavy load vehicles on the autobahn since 1995.	Most expressways are toll roads.	There are some toll roads.	There are some toll roads.	All expressways are toll roads.

All of the expressways in Japan are toll roads. In addition, tolls are sometimes employed for the construction of public roads, tunnels and bridges. This system was introduced to speed up roadway improvements and has facilitated the construction of the expressways which have become the nation's lifeline. There are no subsidies; costs are covered totally by usage fees which are collected on a nationwide basis and pooled in a single account.

Although there are currently few toll roads in the United Kingdom and United States, this situation is under review. In the United Kingdom, "New Roads by New Means (1989)" proposes using the know-how of the private sector to construct, maintain and procure funds to speed up the improvement of the highway system. In the U.S.A. the Intermodal Surface Transportation Efficiency Act (ISTEA) is giving much consideration to the promotion of toll road projects.

Germany has no toll roads but since 1995 heavy load-bearing vehicles have been charged a fee on autobahns. In addition they have a unique system in the ÖFFA (Gesellshaft Für Öffentliche Arebetien) which procures construction funds for the autobahn by issuing bonds which are reimbursed from income derived from special fund revenues such as taxes on petroleum. Therefore, the burden is substantially that of the users.

Where toll road systems have been used they have all been successful in promoting roadway improvement. However, if improvements were only undertaken where a profit could be generated then many areas would be disadvantaged; taken in isolation tolls are of limited value.

2.3.3 Public transportation subsidies

The greatest diversity on national attitudes towards transport subsidization is evident in public transport. In order to make a cross-country comparison subsidies are classified as: urban transport, inter-city transport and the construction of basic infrastructure.

(1) Subsidies for the operation of urban transportation
Users of urban public transport pay fares but revenue rarely covers costs, especially where policies exist to encourage a modal shift or to give preferential treatment to the disadvantaged. Japan is the only country which does not subsidize its urban transport.

In Germany the main arguments for artificially holding down fares are: 1) to ensure affordable mobility; 2) to guarantee mobility to the physically disadvantaged; and 3) to rationalize the whole system by encouraging a shift from individual to public transport. The second and third points affect the need for financial resources for roadways and are thus grounds for subsidies from automobile-related tax revenues.

In the United States the role of subsidies for urban public transportation is to relieve road congestion, reduce air pollution and conserve energy. Urban railways receive subsidies from general funds and from ISTEA, which includes funding from such programs as the Surface Transportation Program (STP) and the Mass Transit Program (MTP). The Highway Trust Fund provides subsidies to local governments who own and operate urban railways.

(2) Subsidies for operation of inter-city transportation
There are comparatively few subsidies for the operation of inter-city transportation though there are some grants of this type in France and the United States.

In France financial aid is considered essential for the development of railway transportation projects, and in the United States subsidies are granted to inter-city railways (Amtrak) with the aims of relieving road and air-traffic congestion, and reducing environmental damage.

In the United Kingdom the Public Service Obligation (PSO) grant, a general grant for non-profit transportation, was started in 1974. However, the PSO grant for inter-city projects was eliminated in 1988 and for the Network Southeast project in 1992. Neither the Trainload Freight or Railfrieght Distribution (RfD) receives government grants.

In Japan, with the exception of a very special example, subsidies for operating expenses are not granted for either inter-city or urban transportation, perhaps because there are many private railways making comparatively good profits.

(3) Subsidies for basic infrastructure

All countries invest public money in the basic transportation infrastructure. In Germany, prior to railway reform, the federal railway (DB) received subsidies from general funds. The federal government has financial responsibility for the newly formed Deutsche Eisenbahn Aktien Gesellshaft (DEAG) and, since 1967, a proportion of petroleum tax revenue is ear-marked for the construction of urban railways, the U-bahn and the S-bahn.

In the United States public funds can be invested in the construction of both urban and inter-city railway projects. The MTP enables the Highway Trust Fund to subsidize the upgrading of existing passenger railway lines for high speed and is providing funds for the development of a linear motorcar line. However, the freight railroads essentially do not receive subsidization from any level of government.

In Japan, subsidies are provided for the construction of railways politically defined as necessary such as urban subway lines and the Shinkansen (superexpress inter-city trains) lines. However, decision-making within the railway companies is respected and only investment is subsidized. Traditionally railways are not seen as part of the public infrastructure and it is generally assumed that railway projects are profit oriented.

2.3.4 Airport improvement systems

France, the United States and Japan all have specific funds for airport improvement. In Japan both decision-making and financing are comprehensively performed at the national level. A special central government account for airport improvements receives landing fees and aircraft fuel taxation revenue.

The burden rate for expenditure on airport improvement is determined in accordance with an airport improvement act of 1956, and with the exception of two international airports (New Tokyo and Kansai), airports are improved without reference to income and expenditure of each airport. The state's involvement in airport investment is the result of a perceived need to bring air transport up to international standards.

In the United States, the federal government established the Airport and Airway Trust Fund (AATF) which accumulates specific funds through ticketing tax, aviation freight tax, departure tax and aviation fuel tax. The National Plan of Integrated Airport Systems (NPIAS) has a grant program based on the Airport Improvement Program (AIP) in which federal subsidies are granted to airports. These subsidies generally cover one-third of airport improvement

Table 2.3. Airport improvement systems.

	Germany	France	U.K.	U.S.A.	Japan
Nationwide Plan	Not existing	Not existing	Not existing	NPIAS	Five year Plan (seventh)
Subsidizing Systems from national government	No direct subsidies. (Only investment to some airports)	Existing (Not for the ADP)	Not existing	Existing	Existing
Cross – Subsidizing Systems	Not existing	Not existing (Existing in the ADP airports)	Not existing (Existing in the BAAplc airports)	Existing (under the Airport Improvement Program)	Existing (under the Airport Special Accounts)

Notes: ADP: Aéroports de Paris
BAAplc: the British Airports Authority Public Limited Company
NPIAS: the National Plan of Integrated Airport Systems

expenditures but depend on the scale of the airport. At local airports, where independent operations are difficult, the federal government may provide as much as 90% of the investment costs, whereas a large-scale airport with many users may receive only 20%. Basically, there is no federal grant system for airport authorities management or operating expenses which must be covered by landing fees, stopover fees, rental fees for terminal buildings, income from concessions and parking, etc. Deficit supplements are covered by the general accounts of the local governments concerned.

In France central government has a "Budget Annexe de l'Aviation Civile: BAAA", a civil aviation special account. With the exception of the airports in Paris, which are under the control of the Paris Airport Public Corporation (Aéroport de Paris: ADP), construction is subsidized but operations are not. Ideally construction, management and operations should be self-supporting so there is no generalized national improvement plan. Capital investment for such projects as expansions requires that an airport authority evaluate the investment in the light of expected returns.

There are no specific funds for airport improvement in either the United Kingdom or Germany. The United Kingdom's Airport White Paper adopts the principle of economically independent, self-supporting companies to run the airports. As a result there are no special accounts and no national cross subsidization between airports (with the exception of airports under the authority of the British Airport Authority Public Limited Company (BAAplc) which are subsidized from BAAplc profits). Aircraft fuels are taxed but the revenue is paid directly into treasury funds.

Germany has neither a specific fund nor a unified, national airport improvement program. This is related to the fact that there is little new investment other than for a new airport in Munich. The international airports are owned and managed by private companies and limited corporations, the federal government participates only as a financier and there are no subsidies from the federal government for airport construction. The companies owning and managing the airports construct runways, taxiways, terminals, parking lots and so on through airport usage fees and debt. Operations are not subsidized.

Despite the range of methods used to fund transportation improvements, it is notable that in general the need for railway improvements is judged within a wider transport policy framework, whereas airports are judged by commercial factors such as profit levels and return on investment.

2.4 INDIRECT FUNDING SYSTEMS

These are systems that take account of the externalities experienced by most transport systems.

2.4.1 Automobile users' taxes and public transportation

First, a significant trend in recent years has been the tendency to utilize taxes collected from automobile users to subsidize public transport. In Germany and the United States especially, it is accepted that a modal shift is necessary to protect the environment, but in Germany there is an additional concern to provide public transport for people currently using private cars who may not be able to do so in the future due to age, accident or sickness.

The enabling legislation in the United States is the Intermodal Surface Transportation Efficiency Act of 1991 (ISTEA'91) .

In Japan there is no cross-subsidization between car users and public transport. However, subsidies have been granted for the basic facilities of the

Table 2.4. Contributors for financial resources.

	Transportation Owner / Operator		User	Population	Specific beneficiaries: Private enterprises, Land owners	Government		Investment	Contributors
	In	Out	Out	Out	Out	In	Out		
A Charges / Fees	Charges	Investment	Charges					Transportation Owner / Operator	Users
B Users' Taxes		Investment	Taxes			Taxes	Investment	Government	Users
C Users' Taxes	Subsidies	Investment	Taxes			Taxes	Subsidies	Transportation Owner / Operator	Users
D General Taxes				Taxes		Taxes	Investment	Government	Population
E General Taxes	Subsidies	Investment		Taxes		Taxes	Subsidies	Transportation Owner / Operator	Population
F Indirect Charges					Charges	Charges	Investment	Government	Specific Beneficiaries
G Indirect Charges	Subsidies	Investment			Charges	Charges	Subsidies	Transportation Owner / Operator	Specific Beneficiaries
H Indirect Charges	Charges	Investment			Charges			Transportation Owner / Operator	Specific Beneficiaries
I Debt	Debt / Charges (Future)	Investment / Reimbursement	Charges (Future)					Transportation Owner / Operator	Users (Future)

monorail and AGT, ostensibly because they are a part of the roadway infrastructure.

2.4.2 "Versement de transport" (France)

Fees are collected from offices along the transport corridor and used to subsidize the improvement and operation of public transport. The justification for this charge is that enterprises benefit from improved access, and affordable fares increase labour mobility. A tax of 1.3–2.4% of total wages is collected from offices with more than nine employees in cities with populations over 30,000. The actual rate is set by regional government. The system, which is optional, was introduced in 1971 and has been adopted in all cities with populations over 100,000. Although every country has fixed asset taxes, and there is a business office tax in Japan, only the "Versement de Transport" is hypothecated for transportation improvement.

2.4.3 Return of development profits

Improvements to transportation systems sometimes create capital gains in the form of increased property value in the surrounding area. In theory a "return of development profits" tax could be used for further transportation improvements. There are no large-scale examples of this system as, in practice, it is difficult to evaluate gains.

Private railway companies in Japan manage real estate business and, consequently, retain the return of development profits. This is one merit of privatization. In the United Kingdom private airport companies develop neighbouring land.

It is worth noting three direct systems implemented in the United States: the Impact Fee (IF), the Special Assessment District (SAD) and Tax Increment Financing (TIF).

IF is a system in which the local government declares a development area and then collects a one-off payment from the developers.

SAD is a system in which the local government declares a continuing development area and then collects fees regularly from the owners of fixed asset who benefit from improved public facilities.

TIF is a system whereby the local government issues bonds to improve public facilities. The bonds are redeemed by the revenues caused by the increase of the total fixed asset evaluation accompanying development.

2.4.4 B.O.T.

B.O.T. schemes (Build Operate Transfer), i.e. concessions of road, rail or airport infrastructure projects to private or mixed (public + private) sector operators, are starting to develop quickly.

In Germany, there is one road B.O.T. scheme (Betreibermodelle) in operation: a major tunnel crossway under the harbor of Rostock (the concession has been awarded to a consortium led by a French contractor), and the Federal Transport Ministry has announced plans for 17 further road B.O.T. projects.

In United Kingdom, there are several B.O.T. schemes in progress on the road network. Two have been completed (a toll bridge on the Thames river, another one on the Severn estuary).

This should be noted as a new method utilizing private funds to road construction and operation.

CHAPTER 3

TRANSPORTATION POLICIES AND FINANCIAL SYSTEMS

3.1 INTRODUCTION

In this chapter the transportation policies and financial systems of five developed countries are described in detail using the framework established in Chapter 2.

Each country section starts with a description of transport policy and the rationale behind it. We then look at the way this policy is implemented through the development and management of roads, rail, and airports with an emphasis on the method of funding, the balance of public and private sector involvement and the use of long term debt.

3.2 TRANSPORTATION IMPROVEMENT SYSTEMS AND FINANCIAL RESOURCES IN GERMANY

3.2.1 Transportation investment policies

(1) Policy and rationale

The direction of transportation policy in Germany in the 1990s is described in "Verkehrspolitik der 90er Jahre (1990)." There are three basic areas: the construction of a more environmentally sensitive transportation system with emphasis on the railways; improvements in the transportation infrastructure following East-West unification; and a step-by-step progress toward European market integration.

Underlying this policy stance is a belief in the efficacy of market principles and deregulation and a need to cope with current transportation problems caused by unification, in particular, the pressing need to modernize transport in the former East Germany. As a result of the opening up of Eastern Europe, Germany

Table 3.1. Financial resources for transportation improvements in Germany.

	Classification of Financial Resources	Financial Resources	Uses

Roadways	Federal government	·General funds of federal government
		·Portion of petroleum tax (gasoline tax)
		·Public loans (ÖFFA[1] issues.)
			·Expressway improvement
			·Improvement of federal highways
			  (Petroleum tax, under authority of Federal Ministry of Transport)
			·Subsidies for improvement of local roads
			·Improvement of local roads
			  (Petroleum tax, under authority of Federal Ministry of Transport)
			·Grant to ÖFFA (for credit publication)

	Local governments	·General funds of state and local governments ·Subsidies from federal government ·Motor vehicle tax ·Public loans	·Improvement of federal highways ·Improvement of local road (Motor vehicle tax, but allotment percentage is not always 100%.)
	Owners / Operators		
	Debt in above-mentioned	·Public loans (ÖFFA[1] issues.)	·Expressway construction
Railways[2]	Federal government	·General funds in federal government ·Portion of petroleum tax (gasoline tax)	·Basic facilities of urban railway ·Subsidies for capital investment in DB ·Grants to local governments (*U-Bahn*) ·Subsidies for operating expenditures of *S-Bahn* ·Improvement of public urban passenger transportation (Petroleum tax)
	Local governments	·General funds of state and local governments ·Subsidies from federal government	·Basic facilities of urban railway ·Subsidies for operating expenditures of *U-Bahn*
	Owners / Operators	(·Fare income) ·Subsidies from federal government ·Subsidies from state and local governments	·Railway improvement ·Improvement of trunk line (DB, national government burden)
	Debt in above-mentioned		
Airports	Federal government		·Investment to five main airports ·Basic facilities (Aerial aid facilities at international airports) ·Financing without interest for improvement of airports[3]
	Local governments		·Investment to international airports ·Improvement of airports to be managed directly ·Operating expenditures of airports to be managed directly ·Financing without interest for improvement of airports[3]

	Owners / Operators	·Investment from federal government
		·Investment from state and local governments
		·Self funds
		  (Airport rental fees and landing fees)
		·Financing without interest from federal, state and local governments[3]
		·Debt from the private sector
			·Improvement of five main airports
			  (Investment from federal government)
			·Airport improvement
			·Basic facilities
			·Operating expenditures

	Debt in above-mentioned	·Debt from the private sector	·Airport construction

1. ÖFFA : Public company
2. Before DBAG (German Federal Railway Corporation :Deutsche Bahn Aktien Gesellshaft) established (Jan. 1994)
3. Munich International Airport

is expected to become a transit hub between East and West so an efficient transport system is a prime requirement. The Ministry of Transport estimates that the flow of people will increase by approximately 32% and the flow of goods by approximately 25% by the year 2010.

At the same time, a policy that reduces road transport and increases the use of the railways is important from an environmental perspective. The German government aims to reduce carbon dioxide output to one-third or one-fourth the current level by 2005, and as a result, investment in the transport is shifting.

(2) Master plan and principle of burden

Under German Basic Law major transportation routes belong to the federal treasury and thus federal government has responsibility for improving roadways and railways (Basic Law, Article 87 of the Constitution, revised 1993). Strategies described in the German Federal Transport Way Plan (Deutsche Bundesverkehrswegeplan) take a unified approach to the construction of long-range roadways, railways, inland waterways (canals), and airports by the Federal Ministry of Transport with the financial burden shared by federal and local governments.

In contrast to most other countries little emphasis is placed on traffic demand and profit. For example, the 1985 plan's aim is to develop local communities, establish equal living conditions in all areas (which the constitution demands), and maintain economic growth and employment. This means that the infrastructure must be created before increased demand can be measured.

The first post-unification plan, covering the period from 1992 to 2010, will cost approximately DM 280 billion. State funding will only cover about two-thirds of this so although most "top priority projects" will be implemented using the federal budget as in the past, the possibility of utilizing private funds is being examined. In one model currently being considered, for example, private fund procurement companies would underwrite the debt with federal government refunding it as a long-term lease fee.

3.2.2 Roadways

(1) Plans and systems

The Federal Transport Way Plan (Bundesverkehrswegeplan), the master plan for all transportation, is also the strategic plan for roadways. The 1992 plan specifies the construction of 2,000 km of autobahn, 5,000 km of national highways and the renewal and reconstruction of 3,000 km of other roads (all figures approximate). Behind this plan is social reform following the re-unification of

Germany and the integration of Europe which brings new transportation demands. However, since the emphasis in this policy is on federal railway investments, expenditure for federal roadways have decreased relative to the 1985 plan.

Traditionally all roadways, including the autobahn, were constructed to provide free passage. In recent years, however, the use of toll fees on the autobahn has been discussed, and a toll on trucks has been in place since January 1995. These changes are intended to bring Germany into line with France and Italy, where toll road systems are already in place, as well helping to cope with the financial burden of construction in the former East Germany. The adoption of a toll system is seen as a significant change from the past.

(2) Division and improvement authorities
The roadway system is divided as follows:

–Federal long-range roads (Bundesfernstraßen)

Federal autobahn (Interstate Highway (Bundesautobahn))

Federal roads (Bundesstraßen)

–State roads, district roads, city and town roads

State roads (Landesstraßen)

Districts roads (Kreisstraßen)

Local roads (Gemeindestraßen)

Urban roads (Roadside and automobile roads)

The federal long-range roads (Bundesfernstrassen) form a country-wide network which is divided into autobahn and federal roads. The relevant legislation is the Federal Long-Range Road Act.

The improvement authorities for roadways in Germany are the Federal Ministry of Transport (Bundesministerium für Verkehr), the states (Land), the districts (Kreis), and the cities and towns (Gemeinde). There are no private companies or public corporations. The Federal Ministry of Transport assumes responsibility for all expenses for federal long-range roads, but has no implementation organization. The ministry only studies, plans, and distributes budgets. Actual construction and management are entrusted to the states (Land) which have the authority to plan, construct and manage state roads. In addition, the states constructs and manages federal long-range roads on behalf of the federation. Each state has ministries that are equivalent to the Ministry of Transport and which cooperate with each other on road projects. The districts (Kreis) and cities and towns (Gemeinde) construct and manage their own roads.

(3) Financial resources

The financial resources behind German transportation are entirely public, since there are no other authorities for road construction, but a large portion of the federation's financial resources is derived from a petroleum tax. Specific funds for roadway improvement were introduced by the Federal Long-Range Road Act (1953) and the Transport Finance Act (Verkehrsfinanzgesetz) (1955). The 1960 Road Construction Finance Act established a system in which 50% of the petroleum tax revenues were allocated to federal road construction but in 1963, as a result of declining economic conditions and expanding military and welfare expenditures, this proportion was reduced to 45% with the remainder going into general funds. Since 1967, hypothecated tax revenues have been made available for other transportation policies.

The ÖFFA, a fund-procuring organization established by the Transport Finance Act of 1955, operates under commercial law to procure funds for the purpose of promoting expressway (autobahn) construction. This bureau issues bonds, then later buys them back using revenue from specific funds. The federal government supplies the ÖFFA with subsidies and ÖFFA issues bonds for fund procurement backed by a governmental guarantee.

The systems mentioned above have become the financial foundation for the construction of roadways in Germany but in addition local government is able to direct a portion of the state-imposed automobile owner's tax towards specific funds. The percentage varies from state to state with a few states investing the full amount. Other methods of generating financial resources at the state level have included general tax revenues, federal subsidies and public debt.

3.2.3 Railways

(1) Plans and systems

Two factors influencing Germany's railway development are East-West unification and environmental problems. Approximately 4,000 km of high-speed railway network has been upgraded since 1985. Out of a total of DM 34 billion, DM 15.56 billion went into the construction of new lines and 10% of the total was set aside for environmental improvements. This network is now part of the European high-speed railway network approved by the EC Transport Committee in 1991.

The Unification Transport Project (Verkehrsprojekte Einheit) of 1991 resulted in a substantial increase in funds going to the new federal areas; in particular, approximately DM 29 billion was targeted for new line construction and the upgrading of approximately 2,000 km of line extending radially from

Berlin. Additionally, the 1992 Federal Transport Way Plan (BVWP-92) high-lighted priority projects to be implemented by 2010, for which DM 108 billion (49%) out of a total of DM 222 billion goes to the railways. This exceeds the amount allocated to roadway improvement. The shift to railways as an environmental measure has become a significant aspect of urban transportation.

Funding facilities for urban railways such as line beds, tunnels, bridges, station buildings, etc. and public transportation, U-bahn, federal railways and S-bahn are considered as urban public facilities and are constructed utilizing subsidies from the federation and the states.

(2) Division and improvement authorities

Germany was re-unified on October 3, 1990. The two national railways, the German Federal Railway (Deutsche Bundesbahn: DB) and the East German National Railway (Deutsche Reichsbahn: DR), were united in 1994 to form the independent German Railway Corporation (Deutsche Bahn Aktien Gesellshaft: DBAG).

The keys to railway reform are the establishment of independent (Unabhängigkeit) railway companies, the separation of infrastructure and operations (Trennung Fahrweg und Transport), the stability of railway financing (Sanierung der Finanzstrukturen), the release of rail infrastructure to third parties and the collection of track lease fees (Eisenbahninfrastruktur Nutzung gegen Entgelt und Zulassung Dritter). The concept of separating infrastructure and operations is interesting. Although this is not unique to Germany, it is a particularly good example.

Of the urban railways, the U-bahn is operated by local government, with some of the other lines operated by private railway companies.

(3) Financial resources

Before the reform of the railway system, federal government general funds were allocated to the federal railway (DB) in the form of capital investment subsidies. In particular trunk lines, both new and existing line improvements were classified as public works projects as set out in the Federal Transport Way Plan.

In the newly established DBAG, the track division has the role of investing capital in railway improvement. The DBAG procures funds for track investment, but if there is a short-fall, financing from the federal government is sought. Reimbursement of this finance is related to depreciation.

Urban railways are constructed with grants received from both federal and local governments. The federal government covers 60% of U-bahn costs from petroleum tax revenue with the remaining 40% coming from states' and cities' general funds. For the S-bahn, the rate of burden is determined by a contract between the railway company and states. Operating expenses are subsidized from general funds: the federal government for the S-bahn, the states for the U-bahn.

In 1993, it was decided that:

1) The responsibility for urban railway passenger transportation would be transferred from the federal government to the states.

2) The states would receive financial support for the maintenance of urban railway passenger transportation from petroleum tax revenue and a fund created by the Urban Transportation Improvement Grant Act (Gemeindeverkehrsfinanzierungsgesetz).

3) The petroleum tax is a federal tax but it is written in Article 106a of the Basic Law that a proportion shall be given to the states for supporting urban transportation.

This was implemented in 1996 thus protecting urban railways through the use of petroleum tax revenues.

3.2.4 Airports

(1) Plan and systems
The federal government has no nationwide unified airport improvement plan; the only large-scale construction in recent years is a new airport in Munich.

(2) Division and improvement authorities
The international airports are owned and managed by private companies and corporations. The federal government has only participated as a financier for the Munich, Frankfurt, Berlin, Hamburg and Cologne/Bonn airports. States and cities act as financiers to other airports.

The federal government constructs the aerial support facilities (radio facilities, weather facilities, air-traffic control), but the companies that own and manage the airports are responsible for constructing runways, taxiways, terminals, parking lots, etc.

(3) Financial resources
Each airport company secures its own financial resources for construction, maintenance and management through airport usage fees and debt. The federal

government supplies no subsidies for construction other than the international airports mentioned above. In addition to owning and managing local airports, local governments also invest and provide no-interest financing to international airports.

Frankfurt am Main International Airport, the center of Germany's air network, is operated and managed by the Frankfurt Airport Company (Flughafen Frankfult/Main AG: FAG) which was established in 1948. The shareholder investment ratio in the company is 25.9% by the federal government, 45.2% by Hessen State and 28.9% by the city of Frankfurt.

In recent years, the largest amount invested has been for the Munich International Airport. The authority is Munich Airport Company Limited (Flughafen München GmbH). For this project the investment ratio is: federal government, 26%: Bayern State, 51%; and city of Munich, 23%. Approximately 75% of the company's total income is related to aerial operations with the remaining 25% derived from terminal building rent. Although there are no government grants, public sector investment can be looked on as a type of subsidy. Construction expenditures total DM 8.5 billion (approximately U.S.$ 6 billion) of which DM 2.5 billion is zero-interest finance from the federal government, states and cities. This can be broken down into DM 0.8 billion stock capitalization and capital surplus, DM 1.7 billion internal reserves, DM 2 billion market interest debt and DM 1.5 billion from the leasing of hangars and catering facilities. For related facilities, the FMG bears one-third of road and water supply costs, 40% of the burden for the construction of 16 km of S-bahn and the station building, with the federal government responsible for the remainder.

For both of the airports mentioned above, the operation costs are covered by revenue from airport lease fees and debt. The federal government never provides subsidies, any deficit has to be covered by debt from the general financial market.

3.3 TRANSPORTATION IMPROVEMENT SYSTEMS AND FINANCIAL RESOURCES IN FRANCE

3.3.1 Transportation investment policies

(1) Policy and rationale

The Basic Law for Domestic Transportation (Loi d'oreintation des transports intériuers: LOTI) was established in 1982 as the basis of a unified

Table 3.2. Financial resources for transportation improvements in France.

	Classification of Financial Resources	Financial Resources	Uses
Roadways	Central government	·General funds of central government	·Improvement of national highways ·Improvement of local roads (under authority of Ministry of Domestic) ·Improvement of free expressways ·Investment to toll expressways (provided from ADF[1] since 1987)
	Local governments	·General funds in local governments	·Improvement of national highways ·Improvement of local roads ·Improvement of free expressways ·Investment to patent approval expressways ·Donation to toll expressways
	Owners / Operators	·Road bonds, Government guarantee bonds, Non guarantee bonds ·Investment to SEM[2] (ADF), Donation from local governments, Investment from local governments	·Improvement of toll expressways
	Debt in above-mentioned	·Public bonds	·Construction of toll expressways
Railways	Central government	·General funds of central government	·Grants for basic facilities expenditures, Grants for local passenger transportation, Special operating subsidies to SNCF[3] ·Capital grants, Compensation for compulsion discount fare to SNCF(urban) and RAPT[4]
	Local governments	·General funds of local governments ·*Versement de Transport*	·Capital grants, Compensation for compulsion discount fare to SNCF(urban) and RAPT ·Long-term low interest loans
	Owners / Operators	·Subsidies from central government ·Subsidies from local governments ·Debt, Bonds, Long-term low interest loans	·Railway improvement
	Debt in above-mentioned		·Railway construction
Airports	Central government	·Fees and taxes collected from airways and passengers (BAAC[5])	·Grants for local airport improvement by BAAC ·Improvement of basic facilities of local airports
	Local governments		·Capital and operating grants for local airport ·Improvement of basic facilities of local airports ·Improvement of airport operating directly
	Owners / Operators	·Subsidies from central government ·Subsidies from local governments ·Self funds by Airport rental fees and so on ·Debt from the private sector	·Airport improvement
	Debt in above-mentioned	·Debt from the private sector	·Airport construction

1. ADF : *Autoroute de France* (French Expressway Organization)
2. SEM : *Société d'Economie Mixte* (Mixed-economy companies)
3. SNCF : *Socoété Nationale des Chemins de Fer Français* (French National Railway)
4. RAPT : *Régie Autonome des Transport Parosien* (Paris Transport Corporation)
5. BAAC : *Budget Annexe de l'Aviation Civil* (Private Sector Aviation Special Account)

transportation policy covering railways, highways, inland waterways and air transport. The following points are characteristic of the policy:

1) Priority for public transport.

2) A new set of rights, the "Transport Rights (droit au transport)," unique amongst capitalist countries.

3) Decentralization of settlements and of transportation policy.

4) Clarification of the key concepts of public service (service public), social expenditures (coûts sociaux) and transportation systems (systéme des transports).

5) Clarification of the role and responsibility of central government in the maintenance, construction and funding of public transport.

This last point is especially important and is worth expanding. It covers: 1) the construction, maintenance and management of fundamental transport facilities (infrastructure) and its safe provision; 2) the regulation and supervision of activities related to transportation; 3) the implementation of a priority public transportation policy for passenger transport; and 4) the implementation of a research programme on the successes and failures of the transportation system, as well as the preparation of statistical data.

All these state activities should be founded on the Transport Rights, namely: 1) the right of mobility to all users; 2) the liberty to choose one's mode of transportation; 3) the right of users to transport themselves personally or to entrust it to others; and 4) the right of users to receive information regarding transportation modes and methods of usage.

Users' rights underpin French transport policies in contrast to Germany where the starting point is public responsibility. But nevertheless, policies in both countries tend in the same direction in practice.

(2) Master plan and principle of burden
During France's period of rapid economic growth, policy was designed to correct "unconditional concentration on the Paris metropolitan area". After the oil shock the emphasis shifted to "expansion of employment in the region" and, in recent years, "the adaptation to an integrated EC market".

"State-Region Development Contracts" are the way in which the financial burden of the transport infrastructure plan (schémadirecteur) is distributed. This mechanism was introduced in the 9th Economic Plan (1984–88). The infrastructure plan is prepared jointly by the governor, who is prefecture representative, and an assembly chairman representing the administrative power. After a preliminary examination by the National Land Improvement Agency (Délégation a' l'Aménagement du Territoire et a' l'Actionm Régionale: DATAR), the minister for national land improvement (Comité Interministeriel de l'Aménagement du Territoire: CIAT) approves the plan.

3.3.2 Roadways

(1) Plans and systems

The first 5-year plan for roadway improvement was agreed in 1951 and 5-year plans have been utilized ever since. In parallel with the 9th 5-year plan (1984–88), a long-range (15-year) plan was drafted. The "National Road Master Plan" which was finally agreed in 1992, included improvements to 37,000 km of trunk road network, including 12,000 km of inter-urban expressways.

Although road construction in France has been slow compared to other European countries, the construction of expressways has been accelerated by the introduction of the toll road system. This was brought into being by the Expressways Act of 1955 which established toll road companies in 1956. The first toll expressway was opened in 1961. Most of the inter-urban expressways currently operate tolls. However, toll fees are not charged to users of urban expressways such as the Paris loop road, a portion of the inter-urban expressways, nor where there are no alternative routes.

The construction and management of the toll roads are entrusted to mixed-economy (Sociéte d'Economie Mixte: SEM) or semi-public companies. Most of the capital of the SEMs is managed by national organizations and neighbouring local governments.

The specific fund for roadway improvement, the Road Investment Special Fund (Funds Spécial d'Investissement Routier) was established in 1951 and funded by revenue from fuel taxes. It was eliminated in 1981.

(2) Division and improvement authorities

National highways are classified in the "National Road Master Plan" (Le Shéma Directeur Routier National) as follows:

–Expressways (Motorways, Autoroute),

–Semiexpressways making up the expressway network (Links ensuring completion of motorway network, Liaisons Assurant la Continuite du Réseau Autoroutier: LACRA),

–Major trunk roads for national land development (Other major land development highways, Grandes Liaisons d'Aménagement du Teritoire: GLAT), and

–Other national highways (Autres Routes Nationales).

Some expressway (Autoroute) and LACRA improvement projects are handled directly by the Ministry of Public Works, Housing, Transportation and Tourism (MPW), and others are entrusted to the SEM or other approved companies, namely, COFIROUTE or Monte Blanc Road Tunnel.

MPW take charge of improvements to free expressways, but construction bureaus of the prefectures (Direction Departmental d'Equipment) are in charge of construction and management in their own areas.

The GLAT is structurally the same as other national highways but it was given as special name to designate its high priority in the national highway network. This, too, comes under the aegis of the MPW, but actual construction and management is through the construction offices of local agencies.

In addition to the above, there are prefectural roads (Chemins Departementaux) for which the prefectures are the maintenance managers and local pavement roads (cities and towns road: Voise Communales) for which the cities and towns (commune) are the maintenance managers. Each prefecture is responsible for the construction and management of prefectural roads through the construction offices of local agencies.

(3) Financial resources

Automobile users are subject to a multitude of taxes; registration tax, drivers license tax, value-added tax, automobile tax, axle tax, special tax for corporate motorcars and fuel tax, but there are no specific funds for roadway improvement so all national highways and local roads, excluding toll expressways, are improved using general funds. Central government shoulders 100% of the burden for inter-urban highways apart from the sections which enter or by-pass urban areas which are jointly funded by both central and local government. The way in which the burden is shared out is decided through negotiations between the two parties.

Budget planning is based on the Long-Range Plan for Road Network Improvement: each area has a five-year plan and each year the budget is implemented according to this plan.

The construction of toll expressways is entrusted to the mixed-economy companies (SEM). The financial resources of the SEM comes from road bonds issued by the companies, guaranteed government bonds, non-guaranteed bonds, prepayment from the French Expressway Organization (Autoroute de France: ADF) and investment from local governments. The ADF was established in 1982 and it has contributed capital to SEM since 1987. The prepayment system is used to adjust finances among the mixed-economy companies and to correct for differences in charges between regions.

3.3.3 Railways

(1) Plans and systems

The 1992 plan for inter-urban railways is called the Basic High-Speed Railway Plan (Le project de schema directeur national des liasons ferroviaires a grande vitesse). It calls for improvement of 4,700 km of railways, 16 TGV routes (including railways in operation and under construction). The TGV improvement plan aroused a great deal of public interest in the cities as it required considerable compulsory purchase of land.

The basis of future plans for TGV are: consistency with the planned high-speed railway network of Europe as a whole thereby contributing to the integration of Europe; creation of fast links between urban regions in France without having to go via Paris; a combination of railway and expressway for freight; congestion reduction on expressways and at airports by the use of an environmentally sensitive transportation system. The construction of new lines is expected to continue until 2015. Although it is quite hard to make profits on any route, other than the southeastern line, measures are underway for aggressive improvement in the railway infrastructure.

(2) Division and improvement authorities

The French National Railway (Sociéte Nationale des Chemins de Fer Français: SNCF), established in 1938, is the authority for the entire national railway network. For urban transportation, there are many public railways such as the Paris Transport Corporation (Régie Autonome des Transports Paroisens: RATP) and a few private railways. SNCF was a mixed company in the form of a joint-stock company which was jointly managed by central government and private railway companies until 1982, thought the private companies have little say in its operation. With the establishment of the Basic Law for Domestic Transportation in 1982 it was changed into an "industrial and commercial public facility (Establissement Public Industriel et Commercial)" for which the government provides all the investment. It enters into contracts with central government and operates independently.

Industrial and commercial public facilities are a form of operation for which the principle of public law is applied to public works and the principle of private law is applied to commercial activities. The Paris Transport Corporation (Régie Autonome de Transport Parosiens: RATP) and French electric power and gas utilities are further examples of this type of organization.

Most urban transportation comes under RATP which was established in 1948 as a transportation project authority by central government and the local governments of the Ile de France area. It operates subways (199 km, 15 routes of METRO) and buses (2,900 km, 269 routes). It also operates the high-speed suburb railway (Réseau Express Régional: RER, 103 km, 2 routes) in cooperation with SNCF.

The STP (Paris Transportation Union) is the umbrella organization for co-ordinating transport policy. It comprises central government (Ministry of Finance, Ministry of Transport), the city of Paris and seven prefectures in Ile de France.

(3) Financial resources

Government grants are considered to be an essential prerequisite for railway development and since the Railway Construction Act of 1842, France has adopted the principle that the public shall bear the burden for the infrastructure and constructors shall bear the burden for tracks and rolling stock.

The financial resources for investment in SNCF are general funds from the central government and investment funds of SNCF itself (inner funds, debt, bonds). Central government provides grants to compensate for social discounts on passengers fares. Each grant application is made according to a set of prescribed instructions, and successful applications result in a contract between the central government and SNCF. When a construction plan and renewal project is implemented, an internal rate of return, which includes socio-economic effects, is calculated. The required return is more than 8–9%.

The grant the for urban railways of the Paris Metropolitan area is submitted by local assemblies (including the Paris City Council), and is set by contracts between central government and regional areas after approval by Cabinet. The burden rate in the case of new construction is generally central government 40%, regional areas 40% and constructors 20%.

Investment grants come from central government, the regions, prefectures and cities; regions provide long-term, low-interest loans.

The "Versement de Transport" provides subsidies to supplement commuter discount fares, and central and local governments provide subsidies to supplement social discounts. There are no grants from fuel taxes as there is no specific funds system. The "Special Fund for Large-Scale Project (Fonds Spécial de Grands Travaux: FSGT)," which was established in 1982, imposed a new tax on fuel for the reimbursement of bonds but was eliminated in 1986.

3.3.4 Airports

(1) Plan and systems

There is no country-wide airport improvement plan so individual schemes must be judged in relation to anticipated demand. The airport network is complete, there is no need for more airports, and growth in demand can be accommodated by expanding capacity through increasing the number of runways. Central government has responsibility for the formation of the airport network and has power to control the number of runways and capacity of passenger terminals in each airport. In practice it does little more than approve development plans as few subsidies are provided.

Since airport construction, management and operation is profit-dependent, each operation authority generates income through airport lease fees, etc., and carries out its own capital investment and operations. In practice only the Paris Public Airport Corporation (Aéroport de Paris: ADP) and 10–15 local airports have sufficient independent profits to manage without help from the public sector. Most of the other local airports receive grants from local governments. The central government does not provide grants for operations.

(2) Division and improvement authorities

Basically, central government constructs airports and the local chamber of commerce operates them, but there has been little new airport construction in recent years. The 14 airports in the vicinity of Paris, including Charles de Gaulle Airport and Orly Airport, are constructed and operated by the Paris Airport Corporation (Aéroports de Paris: ADP). Out of a total of 306 public airports, central government established 161, local governments 93 and the chamber of commerce 27.

Separate authorities handle management and operational matters, and construction. The head office of Civil Aviation of the France Ministry of Public Works, Housing, Transportation and Tourism supervises the ADP and local chambers of commerce, but it is not directly involved in the construction or operation of airports.

The ADP constructs, manages and operates the 14 airports in the vicinity of Paris. One-third of ADPs board of directors are members of government. Fundamental operational issues such as drawing up investment plans and setting landing fees, are placed under the administration of the government. The airports provide self funding and debt for capital investment and there are no subsidies from the central government.

(3) Financial resources

The central government sets the fees and taxes which are collected from airlines and passengers and places the revenue in the Civil Aviation Special Account (Budget Annexe de l'Aviation Civile: BAAC). Subsidies are provided for airport construction except for those airports which come under the Paris Public Airport Corporation (ADP). There are no subsidies provided by BAAC for operating expenses. But local governments grant subsidies for airport construction in addition to providing grants towards running costs. In addition to the BAAC, operating profits (internal reserves), such as airport lease fees of constructors and debt from the private sector, make up their financial resources.

Subsidies provided for the airport infrastructure are limited to 30% of the required funds. In order to construct new infrastructure, airport managers must procure funds independently by soliciting subsidies from local governments or acquiring debt.

There are charges for landing, lighting, parking, passenger use, fuel supply and airport rental charges, as well as income from lease fees of the airport site, hangar, emporium, offices, etc.

3.4 Transportation improvement systems and financial resources in the United Kingdom

3.4.1 Transportation investment policies

In recent years privatization of all forms of transport has been promoted. British Railways and British Airways were both privatized in the expectation that competition would stimulate increased efficiency. The purpose of these policies is to reduce government subsidy expenditures, to encourage the construction of less expensive transportation and to create a system that can benefit from market expansion within a more integrated EC.

Local transport planning was substantially transferred to the local authorities (County and Metropolitan Councils) by the Local Government Act (1974). Implementation is funded through the Rate Support Grant (RSG) and Transport Supplementary Grant (TSG) and local property taxes. Local governments present their plans in a bid for funding from central government in the form of a comprehensive Transport Policies and Programme (TPP) package. Successful bids result in awards of TSG. The TPP covers the whole range of transportation problems including road construction, public transportation, transportation man-

Table 3.3. Financial resources for transportation improvements in the United Kingdom.

	Classification of Financial Resources	Financial Resources	Uses
Roadways	Central government	·General funds of central government	·Road improvement ·Expressway improvement ·Subsidies to local government
	Local governments	·Rate, Rate Support Grant (RSG), Transport Supplementary Grant (TSG)	·Road improvement ·Expressway improvement
	Owners / Operators	·Toll road fees	·Toll road
	Debt in above-mentioned		
Railways	Central government	·General funds of central government	·Subsidies to local government ·PSO to BR[1] ·Subsidies (LT[2])
	Local governments	·Rate, Rate Support Grant (RSG)	·Railway improvement (Excluding BR and LT)
	Owners / Operators	·Self funds by fare income ·PSO grant, ·Subsidies from central government ·Debt from Minister of Transportation	·Railway improvement
	Debt in above-mentioned	·Debt from Minister of Transportation	·Railway construction (BR and LT[3])
Airports	Central government	·General funds of central government	·Airport improvement ·Operating grants to subsidiaries of CAA[4]
	Local governments	·General funds of local governments	·Airport improvement ·Subsidies to non-profit airports
	Owners / Operators	·Self funds ·Subsidies from CAA ·Subsidies to non-profit airports from local governments ·Government financing, Debt	·Airport operating, development, aerial control ·Capital investment by self funds and debt of BAAplc[5] ·Return of development profit
	Debt in above-mentioned	·Government financing, Debt, Debt from the private sector	·Construction of airports of BAAplc

1. Before privatisation of British Railway Board (April 1, 1994)
2. LT : London Transport
3. Since 1985, all financial resources for investment of LT are covered by self funds and subsidies.
4. CAA : Civil Aviation Authority
5. BAAplc : British Airport Authority Public Limited Company

agement and road maintenance, in a five-year plan with detailed estimates of annual expenditure.

3.4.2 Roadways

(1) Plans and systems

During 1975 increasing awareness of environmental problems plus the oil shock led to changing attitudes towards motorway expansion, but when Mrs. Thatcher came to power in 1979 she put road building at the center of her transport policies.

The central government's transport budget was increased by a factor of 1.64 between 1981 and 1990, whereas the national road budget increased by a factor of 2.68. In "Roads for Prosperity" (1989) the argument for a substantial increase in road investment was based on the need to relieve traffic congestion on trunk roads which was beginning to hinder economic development. The plan set about improvement, expansion and new construction of a trunk road network focusing on 2,700 miles of expressways (4,360 km). "New Roads by New Means" (1989) proposed the introduction of privatization for construction, maintenance and funds procurement in order to quicken the pace of roadway improvement. However, there are few toll roads in the present network as compared to other countries.

(2) Division and improvement authorities

Roadways are categorized as follows:

–Trunk roads (motorways, non-motorway trunk roads);

–Principal roads (local authority motorways, classified principal roads, classified non-principal roads); and

–Other unclassified roads.

The construction and maintenance of trunk roads is the responsibility of the central government. Trunk roads were established under the power of the central government for the purpose of aggressively improving and reorganizing existing roads in an attempt to rationalize land use in the United Kingdom. These roads link major economic centers and have an important tactical role. The trunk roads in England are administered by the Department of Transport (DOT), and those in Wales and Scotland are administered by local road bureaux.

Both principal and unclassified road are constructed and maintained by local governments using grants from the central government. Roads which connect inter-urban and urban roadways are classed as principal roads.

(3) Financial resources

Automobile users' taxes include car registration tax, fuel tax and value-added tax. The United Kingdom does not use a specific funds system for roadway improvement; taxes are processed through the general accounts of the central and local governments. Although a specific funds systems for roadway improvement was introduced by the Development and Road Improvement Fund Act (1909), it was eliminated in 1937. However, 20–30% of the revenue collected from automobile users' taxes is allocated to roadway improvement, partly through the Rate Support Grant (RSG) and Transport Supplementary Grant (TSG). The RSG is a formula-based subsidy covering education, housing, social welfare, sanitation, health and safety, law enforcement, etc., in addition to highways. The TSG is specifically for transport. The local authority draws up a Transport Policy and Programme (TPP) proposal as the basis of a bid for TSG. Funds are not granted for individual projects but for the package as a whole. Previously public transport was included but since 1975 bids have been restricted to roads, bridges and road safety schemes. Local governments also issue local government bonds to create financial resources for roadway improvement.

3.4.3 Railways

(1) Plan and systems

There is no comprehensive national railway plan and there is little investment in new lines. Construction of an Inter-City network, an inter-urban high-speed railway, is underway, but there are many cases where 19th century track is being retained. Compared to other countries, there is little new railway investment.

Recent railway reform has been based on the privatization of British Railway Board (BR), which was separated into infrastructural and operational sections in 1994. Railtrack, which was privatized in 1996, manages the rental of track and station buildings and the allocation of train operation schedules. Passenger transport is split geographically into the regions covered by the old, pre-nationalized companies and franchised out to private sector operators on an 'open access' basis. Freight and support services are franchised separately. This side of the business is overseen by a Franchise Director. A Rail Regulator is responsible for ensuring that rail companies achieve the standards and conditions agreed in their contracts.

The possibility of privatizing BR was explored in "New Opportunities for the Railways—Privatization of British Rail" (July 1992). One suggested method was to float BR on the stock exchange as a single company, but it was felt that the system chosen would stimulate competition in contrast to the monopoly situation that would exist if there was just the British Rail Company.

The New Railways Act was passed in 1993.

(2) Division and improvement authorities

In addition to the national system, there are public urban railways such as London Transport and some private railways. Passenger transportation associations (PTAs) were established in seven major cities (Sheffield, Leeds, Newcastle, Birmingham, Liverpool, Manchester and Glasgow). They have expanded track networks, replaced vehicles, promoted electrification and established new stations.

(3) Financial resources

The National Railway received a Public Service Obligation Grant, which is compensation in the form of accumulated funds for nonprofit passenger transportation introduced by the Railway Act in 1974. This system imposed the obligation (public service provision obligation) to maintain the standard of passenger service of British Railway in 1974. On the other hand, as compensation, the system used accumulated funds to compensate all networks including inter-city railways.

There are no grants from the central government for construction, but most infrastructure costs are appropriated as operation expenses, and costs including capital expenses were dealt with by grants.

Subsidies to inter-city projects were stopped in 1988, and the PSO grant to Network SouthEast was eliminated in 1992.

London Transportation (LT), urban railway, was improved with the help of subsidies from the government to top up their internally generated funds. The central government does provide subsidies including grants for operations and for capital. Also, according to the LT Act, funds can be borrowed from the Minister of Transportation. However, since 1985, all investment funds except self-created funds are covered by subsidies.

3.4.4 Airports

(1) Plans and systems

The U.K. has neither a large-scale airport plan, nor a country-wide improvement plan. However, in order to maintain the international status of London and the international competitiveness of the country's airlines, it is recognized that there is a need to upgrade metropolitan area airports.

Outside the metropolitan area there is little airport investment; most airports in Scotland have to manage without state assistance.

From the end of World War II up until the 1960s, the central government improved existing airports and built new ones. More recently governments have stressed the need for airports to become self-supporting and to operate within a market framework. Since competition is the order of the day, there is no cross subsidization between profitable and unprofitable airports.

(2) Division and improvement authorities

Airports are owned by central government, local authorities and the British Airports Authority Public Limited Company (BAAplc). Some local government airports are managed and operated by the private sector. There are several airports on isolated Scottish islands which are owned by the Civil Aviation Authority (CAA) which operates as a company organization, and Belfast International Airport is operated by a subsidiary of a national limited-stock company.

BAAplc owns and co-ordinates the airports of Heathrow, Gatwick, Stansted, Southampton, Glasgow, Edinburgh and Aberdeen; operational matters are handled by a subsidiary. The British Airport Authority (BAA) was established as a national enterprise in 1966 under the Airport Authority Act of 1965, and privatized as BAAplc by the Airport Act (1986).

(3) Financial resources

The central government only grants subsidies to airports on the isolated highland islands in Scotland which are under the management of the CAA. Airports owned by local governments are constructed and operated using self-generated income from airport rental fees, etc. In the case of nonprofit-making airports, the deficit is made up from the general accounts of the local authority. There are no grants for investment or operations in BAAplc and its financial resources come from self-created revenues and debt.

Although an aircraft fuel tax is collected it is not hypothecated for airport improvement but is placed in the treasury as general funds.

In summary, the sources of income available to airport authorities are: 1) airport rental fees for air transport (landing fees, passenger fees, parking fees, etc.); 2) commercial rental fees (concessions, parking lot fees, etc.); 3) rental fees for the air traffic control facility; and 4) subsidy income (only small local airports with deficits). Airport rental fees are not unified nationally and can be set at the discretion of the operators. Rental fees and commercial services are not subject to special regulations.

Some of BAAplc airports do operate with a deficit so cross subsidization occurs within the company.

3.5 Transportation investment policies in the EC (EU)

(1) Political objectives

Policies of individual European countries clearly relate to the policies of the EC. The 1992 Maastricht Treaty incorporated improvement to the Trans European Network. The EC document "Transportation in a Fast Changing Europe (Transportation 2000 Plus report) (1991)" recognizes that the situation is critical because transportation infrastructure is failing to cope with increasing demand. In particular urban public transportation, roads, airports and access to airports, suffer from insufficient capacity. Political objectives for transportation in EC are to increase the freedom of movement of people, goods and services, to ensure fair competition and to maintain economic growth. It is also believed that transportation policy should promote competition where this does not conflict with other EC policies.

However, constraints on the use of the free market are inevitable in the face of concerns over the environment, energy and safety. There is also concern to enhance international competitiveness. Central to these aspirations is the appropriate burden of transportation costs and the balance of fund procurement between the public and private sectors, as well as aspects of taxation systems and the promotion of technological development. Recent concerns about sustainable development mean that all transportation policies are designed to have minimum impact on the environment by: 1) maximize the use of existing capacity, 2) strict pollution standards, 3) the development of environmentally neutral transportation modes and techniques, 4) the location and development of industrial and commercial activities should be correctly reflected in the transportation costs imposed on the entire society.

(2) Plans and burden principle

The EC has an annual grants budget of 140m ECU to devote to the development of a European-wide transport network. Priority projects are those which eliminate missing links and bottlenecks and which attempt to equalize the level of modernization and technical know-how across countries. Improvement in the high-speed railway network and Combined Transit is especially relevant as a means of relieving road congestion and alleviating environmental problems. There are no subsidies for airports.

The amount of financial support coming from the general accounts of the EC is not so large but adoption by the EC raises a projects priority by the recognition that it is of European interest. Such notification makes it easier to find finances with favorable conditions, and interest may be refunded by European investment.

3.6 TRANSPORTATION IMPROVEMENT SYSTEMS AND FINANCIAL RESOURCES IN THE UNITED STATES

3.6.1 Transportation investment policies

(1) Political objectives

The 1991 Intermodal Surface Transportation Efficiency Act (ISTEA) requires that intermodal transportation systems are to be developed nationwide. The systems must be economically efficient, sound for the environment, energy efficient and strengthen the competitiveness of the United States in the world economy. Core political objectives are: 1) To build a National Highway System (NHS) by selectively investing in interstate highways or important roadways connecting airport and harbors. 2) To allow the states and local governments to make flexible plans for choosing public transportation or road transportation as well as allowing them to implement optimal choices using new techniques for planning and management. 3) To support the development of new technologies for transportation in the 21st century such as the Intelligent Transportation System (ITS) and systems for magnetic levitation (linear motorcar). 4) To realize the ownership of toll roads by the private sector and to ease regulations concerning the use of the federal budget for toll roads. 5) To continue grants for public transport. 6) To release highway funds for bank construction in swamp lands, preservation of wildlife, plants and historical sites, reduction of air pollution, improvement of bicycle paths and walkways, and environmental improvements such as road landscape. These policies are based on a recognition that transportation system development is failing to keep up with increases in demand for automobile and air transport, and even that the ability to maintain appropriate maintenance is questionable. It was also accepted that a modal shift to more environmentally sensitive transport is necessary.

(2) Plans and burden principle

The year before ISTEA was passed, "Moving America ... New Direction, New Opportunities (1990)" showed the direction for the future. Transportation policies should incorporate optimal intermodal mixes of transport for moving people and goods through deregulation. But there is a need to increase the responsibilities and power of the states and local government as well as using the private sector. In particular there is a need to reinforce the "principle of beneficiary burden" by increasing the weight of users' taxes, increasing returns

Table 3.4. Financial resources for transportation improvements in the United States.

	Classification of Financial Resources	Financial Resources	Uses
Roadways	Federal government	·General funds of federal government ·Highway Trust Fund (Specific funds) Fuel tax and so on ·Interest revenue of Highway Trust Fund	·Improvement of direct control roads ·Subsidies to local governments ·Subsidies to state and local governments for toll road improvement ·Exceeding necessary expenditure invests in the public debt
	Local governments	·Subsidies from federal government ·General funds of local governments ·Highway Trust Fund (Specific funds) Fuel tax, Registration tax and so on ·Fees from toll roads ·Public bonds	·Improvement of Interstate and Defense Highway ·Improvement of state roads ·Roadside improvement ·Redemption of public debt ·Improvement toll roads
	Owners / Operators	·Subsidies from federal government ·Funds from the private sector (Contribution from beneficiaries, Return of development profit)	·Improvement toll roads
	Debt in above-mentioned	·Public bond	·Road construction
Railways	Federal government	·General funds of federal government ·Highway Trust Fund (STP[1], MTP[2])	·Subsidies to local government ·Capital and operating grants to Amtrak by general funds ·Capital and operating grants to Urban railway based on ISTEA[3]
	Local governments	·General funds in local governments ·Subsidies from federal government to public transportation based on STP, MTP	·Capital and operating grants to Amtrak by general funds ·Capital and operating grants to Urban railways by subsidies from federal government and general funds
	Owners / Operators	(·Fare income) ·Subsidies from federal government ·Subsidies from local governments ·Burden of private enterprises ·Bond, Return of development profit	·Railway improvement
	Debt in above-mentioned	·Bonds (Urban railway)	·Railway construction
Airports	Federal government	·Airport and Airway Trust Fund (Specific funds) Ticket tax, Aircraft fuel tax, Aerial freight tax ·Interest revenues of AATF	·C.I.Q ·Maintenance and administration of aviation safety facilities ·Subsidies for airport improvement to constructors and local governments

	Local governments	·Subsidies from federal government
		·General funds of local governments
		·State aircraft fuel tax
		·Self funds
		  Landing fees, Airport rental fees
		·Local bonds
		  General funds bond, Revenue bond
			·Airport improvement (as constructors)
			·Maintenance and administration by airport rental fees(as constructors)
			·Subsidies for airport improvement by state aircraft fuel tax
			·Subsidies for operating expenditures by general funds (in the case a the deficit)

	Owners / Operators	·Self fund Landing fees, Airport rental charge ·Subsidies from federal government ·Subsidies from local governments ·Bonds	·Airport improvement ·Maintenance and administration by airport rental fees
	Debt in above-mentioned	·Local bonds ·General fund bonds, Revenue bonds ·Bonds	·Airport construction

1.   STP : Surface Transportation Program
2.   MTP : Mass Transit Program
3.   ISTEA : Intermodal Surface Transportation Efficiency Act of 1991

on development profits, extending toll road systems, utilizing private sector funds more and so on. Also, in order to improve the environment and relieve congestion, it is proposed that roadway and land use plans should be integrated, and that growth management and demand management should be used to co-ordinate transportation improvement and land development.

3.6.2 Roadways

(1) Plans and systems

ISTEA developed the concept of a nationwide trunk road network (the National Highway System: NHS) (ISTEA Article 1006) composed of the most important roadways for interstate movement and national defense, roads that connect other modes of transportation such as harbors and airports, and roads that are indispensable for international commercial activities. The NHS covers about 166,000 miles throughout the entire country.

The NHS is seen as the first step toward building a comprehensive National Transportation System (NTS) for land, sea and air in the 21st century. It is expected that the mobility of people and goods will become safer, faster and more comfortable, and that industrial competitiveness will be strengthened as the result of reducing transportation costs and increasing domestic employment. The annual federal funds for NHS set for the period is more than U.S.$ 3 billion annually. As of now, the network is almost complete with net new routes of less than 2%.

The promotion of toll road projects is also an important point of the reform by ISTEA. This is a policy to promote the improvement of roadways through federal government co-operation with states, local governments and the private sector. The states can entrust the construction and operation of toll roads to public organizations or the private sector and federal grant funds can be utilized.

(2) Division and improvement authorities

Road improvement projects are divided into federal grant projects pro-moted by states and local governments using federal, states and local government funds, toll road projects whether private or public, and projects implemented individually by the states and local governments.

In the past, roads entitled to receive federal grants were divided into four categories: interstate and defense highways, primary roads, urban roads and secondary roads. ISTEA has revised these categories down to two: the national highway system and other roads.

Government transport administration is handled by the Department of Transportation, and the Federal Highway Administration (FHWA) manages road administration. There are nine FHWA Divisional offices across the regions which oversee the grant plan through individual State Highway Departments. It is the State Highway Departments that actually implement the projects. The federal government only manages special roads directly and other roads are managed by the states, counties, cities or towns.

(3) Financial resources

The federal government adopted a specific funds system, the Highway Trust Fund. A large portion of automobile users' taxes such as fuel tax, user tax and tyre tax go into this fund. Funds are also defrayed from general accounts to road improvement projects.

The Highway Trust Fund was established by the Highway Revenue Act in 1956 for the purpose of promoting improvements to interstate and defense highways. Before the fund was introduced, automobile users' taxes collected as general funds were transferred to specific funds for road improvement. Over 70% of the funds were utilized as financial resources for the construction of interstate and defense highways. In addition, on the principle of "pay as you go," it was not permitted to issue public bonds to create future income for reimbursement.

The National Highway Traffic Safety Administration (NHTSA) and the FHWA control road safety projects which are funded by the Highway Trust Fund and from general accounts. Safety plans are drawn up by the states and local governments. The federal government guides the road policies in each state on such issues as improvement priority, decisions for routes and the integration of road standards and technical standards. Federal grants do not cover the whole road works costs.

Many states have established highway trust funds comparable to that of the federal government. Income comes from cities and towns, automobile users' taxes, revenues from toll road fees and the investment of general funds as income in addition to the subsidies provided by the federal government. Expenditures for state roads managed under federal grants, expenditure for other state roads and subsidies to cities and towns are included in the road expenditures of states. ISTEA established the Surface Transportation Program (STP) grant programme to which states and local governments can apply for all roads (including NHS) except collector and local roads. Using this program, it is possible to divert investment to public transportation (buses, commuter railways, etc.). Each state uses 10% of its allocation on safety projects such as the intersections of railways and roads, places 10% in reserves for investing in the improvement of transport quality in terms of environmental preservation, distributes 50% to regions with population of more than 200,000 and other regions according to population, and distributes the remaining 30% to regions of their choice.

Financial resources for toll expressways come from the collection of fees which cover the full costs of each individual route. Currently, there are approximately 5,000 km (interstate highways, 4,300 km (1993)) of toll expressways. The state governments sometimes construct toll roads themselves but, more often, committees comprising state and local government organizations, or public corporations called "authorities," undertake construction. Private companies also construct toll roads. Currently there are about 100 organizations, including states, local governments, authorities (public corporations) and private companies who undertake the work. At present, only a few states have issued bonds for road improvement and there are few organizations where counties and local governments issue bonds.

3.6.3 Railways

(1) Plans and systems

Inter-urban railways are operated by Class I Railroads, which accounts for most of the freight handled, and Amtrak, which is the only passenger railway company. The existing networks are large and there are no concrete plans for new large-scale investment in inter-urban railways. In recent years investment has been directed towards urban and suburban railways in an attempt to relieve congestion during rush hours and on improving the environment. Investment has been made in airport access.

ISTEA introduced the Surface Transportation Program(STP) and Mass Transit Program (MTP) for railway improvement. The STP is mainly a regional grant program for road improvement but it can be diverted toward the improvement of public transportation (buses, commuter railways, etc.). The MTP gives substantial power to states and local governments regarding the choice of public transport, etc. This program makes it possible to receive funds from the Highway Trust Fund to speed up the completion of projects on existing passenger lines and to develop the magnetically levitated railways (linear motorcar), as well as to issue bonds guaranteed by the government for the construction of high-speed railways. In total, the MTP budget is U.S.$ 31.5 billion, of which U.S.$ 18.2 billion comes from the Highway Trust Fund and U.S.$ 13.3 billion from general funds.

(2) Division and improvement authorities

There is no national railway in the United States. The large-scale freight companies are known as Class I Railroads, and Amtrak provides inter-urban passenger transportation. At present, there are 12 freight companies operating as

Class I railways with an annual income of over U.S.$ 253.7 million. They account for only about 2% of the number of railways in the United States but handle approximately 70% of the total operation mileage of all railways, employ approximately 90% of total rail staff and earn approximately 90% of the total freight income.

Amtrak was established in 1970 as a result of the Rail Passenger Act of 1970 and began operations in 1971 with federal government subsidies and private sector funds. When first beginning operations, the company had a rate of income to expenses of around 40 percent, but this has grown to approximately 80% in recent years. It has approximately 25,000 miles (40,225 km) of track and about 500 stations. Train movement is managed by participating railways. Only in the northeastern corridor, between Boston and Washington, did Amtrak buy track, station facilities and rolling stock from Conrail. It has conducted operations directly since 1976.

Railways in the urban areas are constructed and operated by local governments, such as the Los Angeles Mass Transit Authority (LAMTA), Washington Metropolitan Area Transportation Authority (WMATA), New York Metropolitan Transportation Authority (MTA). Some private railways also exist.

(3) Financial resources
Amtrak cannot make profits without support. It receives capital grants to purchase rolling stock and modernize facilities, and grants for operating, mainly from the federal government. The organization also receives financial support from states and local governments based on the Rail Passenger Act of 1970, to provide passenger services requested by the states. The federal government carries the full burden of investment in Amtrak's Northeastern Corridor route (600 or so miles between Washington, D.C. and Boston, Massachusetts), but states are sometimes burdened with a part of the investment for other routes.

Grants for urban railways are provided from general funds and grants based on ISTEA including the STP and MTP which come from the Highway Trust Fund to local governments. On the other hand, local governments individually construct and operate railways in their region, and grant capital and operation subsidies to Amtrak from their general funds. In addition part of the financial resources for urban railways are secured by bond issues and returns of development profits.

3.6.4 Airports

(1) Plans and systems
Airport improvement plans are shown as the National Plan of Integrated Airport systems(NPIAS) of the Airport and Airway Improvement Act (AAIA,

1982). "Subsidy plans" are drawn up by the Airport Improvement Program (AIP). The NPIAS is an improvement plan for public airports which the Federal Aviation Administration (FAA) summarizes every two years. The FAA estimates the level of expenditure to improve those airports needed to provide a country-wide system that will meet passenger demand, as well as support national defense and the postal system.

Airports included in the NPIAS are chosen from applications submitted by states, counties and cities. Judgment is based on expected future demand and its role in the national network. Federal subsidies are granted to airports based on this plan. In NPIAS 1990–1991 the plans for 3,285 existing airports and the construction of 407 new airports were reviewed. Estimates of projects for the 10-year investment plan 1990–1999 total U.S.$ 40.5 billion. Of this amount, it is assumed that one-third will be procured by subsidies and the remaining two-thirds will be procured by local governments in their role as airport authorities. When investment is analyzed by airport type, the proportion taken by large hub airports is the highest (50.5%). Pavements and illumination account for 45.6% and terminal buildings for 14.7%. Investment in the extension of airports and construction of new airports accounts for 69.5%.

The AIP sets the ways for implementing investments in airports based on NPIAS. The basic idea of the AIP is to disperse subsidies to numerous smaller airports while suppressing grants to large airports. It is believed that large airports should re-invest their profits or procure funds from outside. They should also subsidize small, unprofitable airports. New airports require heavy initial investment with a long pay-back period, so there is need for a large public grant at the early stages of construction and then in time funds are recovered from operational income. However, local airports with few users, cannot expect to become profit-making, and projects must be judged on its importance to the aviation network and the provision of basic aviation transportation. In these circumstances continuing subsidy may be required.

(2) Division and improvement authorities

The majority of airports are operated by local governments (city and country), port authorities and airport authorities. States also possess and operate some airports. There are 4,169 public airports and 1,429 private airports. However, the 10 largest airports account for one third of passengers.

The federal government supports financial resources by subsidies for airport improvements. In addition, the FAA has a budget to cover maintenance and expenditure on aviation safety facilities. It grants federal funds for airport improvement to about 3,700 airports included in the NPIAS. The roles of the FAA include: a) determining airport improvement plans for the entire country;

b) granting subsidies to each airport based on airport improvement plans; c) setting design standards and construction specifications for airports; and d) providing air traffic control.

The main roles of the state governments are to determine airport improvement plans and provide subsidies from state fuel taxes. Some states such as Alaska and Hawaii own and operate their own airports. Cities and counties also own and operate airports. In such airports, the cities and counties compile detailed plans for future improvement and then implement them with subsidies from the federal and state governments.

It is the responsibility of each city and county to fund operations and maintenance of their airports, and they are responsible for the burden in the case of a deficit. There are no federal subsidies.

(3) Financial resources

The federal government established the Airport and Airway Trust Fund (AATF) and introduced taxes such as ticket tax, aviation freight tax, departure tax and aviation fuel tax, which are collected for specific funds as laid down in the Airport and Airway Revenue Act (AARA) of 1970. The AATF also has income from interest on deposits in the U.S. Treasury Department. Using these funds, the FAA makes grants based on the NPIAS. In total, one-third of the airport improvement expenditures in the United States is covered by federal subsidies, and local governments (cities, counties) cover the remainder by inner reserves and outside debt (mainly revenue bonds) by independent accounts.

The ratio for subsidies varies according to the scale of the airport. For local airports where independent operations are difficult, the burden of the federal government can be as high as 90%, but the level of subsidies to large-scale profitable airports is approximately 20%. Various bonds are utilized including general obligation bonds, for which refund is made using all available financial resources, and revenue bonds, which are the bonds local governments issue to create funds from the reimbursement of airport income. Additionally, if an airport is in deficit, funds may be provided from the general accounts of local governments.

Improvements carried out by airport authorities utilize financial resources such as federal subsidies, local government subsidies, self-created funds and debt. Revenue bonds such as tax-free local government bonds can be issued for the procurement of investment funds for public purposes to states, local governments, special areas and special organizations as described by Section 103 of the International Revenue Code of 1954. Revenue bonds are only redeemed by income from the airports, and the local authorities who issue them have no obligation to raise other funds for reimbursement.

The source of funding for airport construction at general airports works out as 30% federal subsidies, 35% bonds (revenue bonds) and approximately 30% self-created funds. Operating expenses are covered by landing fees, airplane parking fees, rental fees of terminal buildings and income from concessions and parking lot fees, as federal government has no grant system for these expenses. Supplements for deficits must be covered by the general accounts of the related local governments. When cities and counties own and operate airports, competition for debt may become a problem, but a highly profitable airport generates income for the city or county. Single organizations may procure funds at a lower interest rate than the general account.

As an airport authority is a semi-independent organization, it has the merit that fund procurement costs are low, it has financial independence and income is not diverted to other required funds but remains within the airport. Another merit is that the burden equality can be promoted by creating an organization that includes all of the related local governments.

3.7 TRANSPORTATION IMPROVEMENT SYSTEMS AND FINANCIAL RESOURCES IN JAPAN

3.7.1 Transportation investment policies

(1) Political objectives

The report by the Transportation Policy Council explains the political objectives of improving transportation as follows: 1) to resolve the transportation problems in major cities, 2) to stimulate transportation in local regions and to maintain public transportation, and 3) to contribute to developing a balanced national economy.

Commuter congestion in rail and road are serious in major cities, and the maintenance of public transportation in local regions is problematic. The root of the matter is that population and industries are concentrated in major urban areas. Improvements in transportation are closely related to such fundamental problems of the national settlement patterns.

This matter has become an important theme for the fourth comprehensive national development plan. Settlement dispersal and transport networks are key factors for the 21st century.

Table 3.5. Financial resources for transportation improvements in Japan.

	Classification of Financial Resources	Financial Resources	Uses
Roadways	Central government	·General funds of central government[1] ·Specific funds of central government Gasoline tax, Petroleum gas tax, Motor vehicle weight tax[2]	·Improvement of national highways ·Subsidies to local governments ·Investment to public corporations
	Local governments	·General funds of central government[3][4] ·Specific funds of local government Gas oil delivery tax, Petroleum gas transferred tax ·Subsidies from central government	·Improvement of national highways ·Improvement of local roads ·Investment to public corporations ·Subsidies to public corporations
	Owners / Operators	·Investment from central and local governments ·Subsidies from local governments ·Fees from toll roads ·Road bonds, Treasury investments and loans, Debt	·Improvement of toll roads
	Debt in above-mentioned	·Road bonds, Treasury investments and loans, Debt	·Construction of toll roads
Railways	Central government	·General funds of central government[1] (Railway Development Fund) ·Specific funds of central government Revenues created from selling the Shinkansen (Superexpress)	·Subsidies and financing to public corporations and constructors
	Local governments	·General funds of local government[3]	·Investment and subsidies to constructors
	Owners / Operators	·Subsidies and financing from central government ·Investment from local governments ·Subsidies from local governments ·Fare income, Railway bonds, Debt	·Railway improvement
	Debt in above-mentioned	·Financing from central government ·Railway bonds, Debt	·Railway improvement
Airports	Central government	·General funds of central government[1] ·Specific funds of central government Aircraft fuel tax ·Shares of local governments ·Airport rental fees ·Treasury investments and loans	·Airport improvement ·Basic facilities ·Subsidies to local governments ·Investment to public corporation and private company
	Local governments	·General funds of central government[3] ·Specific funds of local government Aircraft fuel transferred tax ·Subsidies from central government	·Airport improvement ·Basic facilities ·Shares to central government ·Investment to public corporation and private company
	Owners / Operators	·Investment from central and local governments ·Investment and debt from the private sector	·Airport improvement
	Debt in above-mentioned	·Treasury investments and loans ·Debt	·Airport construction

1. Including government bonds in general funds of central government
2. Motor vehicle tonnage tax is general fund. However, we treat it as the specific funds on the character.
3. Including local bonds and Grants of local allocation tax in general funds of local governments
4. Including motor vehicle tax and light motor vehicle tax in general funds of local governments.

On the other hand, in the Basic Plan for Public Investment (1990), the direction of transportation investment is shown from a rather different viewpoint. It recognizes that in Japan, compared to other advanced countries, projects associated directly with the quality of peoples' lives are lagging behind in spite of a high level of public investment. Therefore future investment needs to be focused on improvement of daily use of roadways and subways.

After the rapid advancement of improvements to transportation trunk lines, the network of expressways, Shinkansen (Superexpress), and airports areas are almost complete. The remaining problems are urban transport in the major cities and the need for advanced transport systems in local areas.

(2) Plans and burden principle

The full cost principle, in which usage fees cover all expenditures has become common for expressways, railways and airports. An important characteristic in comparison with other countries is that not only operation costs but also construction costs and even land purchase costs should be covered by usage fees. This basic principle has been followed in the past, and contrasts with all other countries in which the privatization of operating authorities has occurred in recent years. This idea brings conspicuous result in terms of the entire transportation system as the ratio of toll roads is large and public grants for improvements in the railways are small.

In addition the need to introduce long-term debt is emphasized from the view point that the burden of each generation requires adjustment. These principles are not only for inter-city transportation, but for all fields including urban transportation.

3.7.2 Roadways

(1) Plans and systems

Specific funds for highway improvement and toll road systems have become the major pillars supporting road policies.

The specific funds system was founded on two basic ideas: 1) people who gain profits from using the roads pay costs, and 2) people who cause damage to the road pay costs. Presently, a total of eight taxes are collected by central and local governments and directed towards specific roads funds. These funds reached a level of 4.8 trillion yen, and are the main cause of a steady increase in road investment.

As for the other pillar, the toll road system is utilized for all expressways and some general roads, tunnels and bridges. Expressways, which have become the trunk lines, have been especially favored by the toll road system. However, now that most major routes have been completed, attention has shifted to local routes. The 1992 Road Council report expresses the opinion that expenditure cannot be covered solely by the collection of usage fees for expressways constructed in the future. In order to implement the plan for 14,000 km of high-grade trunk roads, which is the present objective, it is recognized that although the toll road system is important, it is necessary to consider the introduction of public grants.

(2) Division and improvement authorities
The division of roadways is as follows:

–National expressways,

–Urban expressways in metropolitan areas,

–Other toll roads,

–General national highways,

–Prefectural roads and

–City, town and village roads.

The national expressways form the main automobile transportation network, connecting important areas of politics, economy and culture. They are constructed by the Japan Highway Public Corporation. The Ministry of Construction issues construction directions to the Japan Highway Public Corporation based on improvement plans.

Toll expressways were extended by approximately 5,677 km in 1995. Plans have been made to extend routes to 11,520 km by the revision of the Major National Land Development Highways Construction Act (1982). Including the 2,500 km of upgraded roads from national highways, a 14,000 km network is planned. In this plan, the objective is to enable people to arrive at interchanges within one hour from all cities, towns and villages.

There are currently urban expressways in the Tokyo metropolitan area and the Kansai urban area. Authorities in charge of construction and maintenance are the Metropolitan Express Public Corporation and the Hanshin Expressway Public Corporation, respectively. Both of these public corporations implement projects dictated by the basic plan provided by the Ministry of Construction.

There are also other toll roads classified as urban expressways in designated cities (Nagoya, Fukuoka and Kitakyushu), the Honshu-Shikoku expressway, and general toll roads which public companies of local governments construct and maintain.

The general national highways form a wide-area transportation system. By 1990 there were 401 routes totaling 44,253 km. The final objective in the "Future Conception of a Trunk Road Network" (1967) was 50,000 km, which is about 4% of the 1.10 million kilometers of all roads.

(3) Financial resources

Gasoline tax and petroleum gas tax are collected by central government for specific funds; and automobile acquisition tax, gas oil delivery tax are collected by local governments for specific funds; and local road tax, petroleum gas tax and automobile weight tax are transferred from central to local government for specific funds.

Although the automobile weight tax goes into the general funds of the central government, 80% of the fund is allotted to road improvement, and since its use is substantively limited, it is usually treated as specific funds.

There is also an automobile tax and a light vehicle tax, which contributes to general funds of the local authority.

Most of the financial resources for toll road construction are some form of debt, e.g. guaranteed government bonds, public enterprise bonds and private placement bonds. The central government grants subsidies to local governments and invests in the four road-related public corporations. The local governments also invest in the public corporations and grant subsidies.

Toll rates are determined under the principles of "reimbursement," "rational charge" and "benefit equivalent." The reimbursement principle is that the reimbursement of expenditures such as all construction, maintenance and management expenditures as well as interest affecting debt should be made by the continuous collection of a fee for a period of time. The rational charge principle is that a rational fee should be charged considering the balance of the burden capability of the users and the fares for other transportation. The benefit equivalent principle is that fees should be charged in a range which does not exceed profits gained from using roads.

3.7.3 Railways

(1) Plans and systems

The Shinkansen (Superexpress) improvement plan, which was published in 1970, is the only nationwide trunk railways plan. The public work projects for roads, airports and harbors have five-year plan, whereas no such plans are prepared for railways neither are there any specific public financial resources for railway improvement.

However, the Railway Development Fund Act of 1991 established the Railway Development Fund on the basis that grants from central government are necessary for the improvement of railways, as large capital investments are needed and there is a long interval before profits can be realized. In the first article of the Railway Development Fund Act it is stated that the purposes of the Railway Development Fund are to promote construction of the Shinkansen (Superexpress), major trunk railways and urban railways, and to grant subsidies to railway constructors. However, it is emphasized that the business policy of the railway constructors must be respected, and that grants simply encourage them to undertake investment For this reason, the objectives of the grant are to promote the policies of the public authorities, such as Shinkansen improvement, congestion reduction in the subways of major cities, overall promotion of residential land development and railway improvement, and modal shifts for environmental reasons.

(2) Division and improvement authorities

The construction and operation authorities of trunk lines are JR companies which were privatized in 1987. Construction is also carried out by the Japan Railway Construction Public Corporation and third sector enterprises.

There are many suburban railways owned by private companies and subways owned by local governments or Teito Rapid Transit Authority. In recent years, some of the new lines are constructed by third sector enterprises, in which local governments and private companies have invested.

In the Railway Project Act (1987) railway construction companies are divided into three types; companies that both own and use track facilities; companies that own track facilities but do not use them and companies that use but do not own track facilities. This Act recognizes that there is a need for third sector enterprises to operate efficiently on railways constructed by local governments, there is a need to separate the operation of transportation business and the construction of infrastructure and that the scale of railway systems constructed by private railway companies has increased.

(3) Financial resources

The central government established the Railway Development Fund by selling the Shinkansen (Superexpress) to JR. This fund is used for grants to improve the Shinkansen, major trunk railways and urban railways. Transfers from general accounts and a special account for industry investment called NTT-B are also included in this fund.

The central government also uses general funds to provide grants to public subways and new town railways. Local governments use general funds to invest in and give grants to public subways. Debt from the central government is used in the case of Teito Rapid Transit Authority and the Japan Railway Construction Public Corporation, while debt from the private sector is used in the case of other authorities. Private railway companies does not receive subsidies other than a preferential tax system.

As for the burden rate for new Shinkansen improvements, 35% is held by the central government, 15% by the local governments and 50% by JR companies of which 60% (30% of all construction expenditures) is in outlays from the Railway Development Fund. The remaining 40% is lent by central government without interest through the Japan Railway Construction Public Corporation. These funds are returned as lease fees which the JR companies pay the Japan Railway Construction Public Corporation when the Shinkansen is up and running. None of the JR companies is required to provide a capital outlay for construction expenditures until the Shinkansen begins operations.

3.7.4 Airports

(1) Plans and systems

Domestic airports have reached the required level and the air network is nearly complete. However, airports in Tokyo and Kansai metropolitan areas are operating at full capacity, even thought a new airport was recently completed in the Kansai area.

It is said that the level of improvement in airport runways in Japan is 70% as compared to western European countries. In comparison to airport improvement in the advanced countries of the world, Japan appears to be somewhat lagging. Furthermore, since the New Tokyo International Airport (Narita airport) is 66 km from the center of the city and is handicapped by the fact that only few domestic flights can use it, there is concern over its status as an eastern Asia hub airport.

Airport improvement plans are comprehensively decided at the national level and reported in a series of seven five-year plans. The sixth five-year plan (1991) identifies the need to promote international communications, to encourage settlement dispersion by promoting international communications in each area and to cater for expected increases in demand.

TO ACHIEVE THESE AIMS, THE SIXTH FIVE-YEAR PLAN PROPOSES TO:

1) Secure the high-standard service level of international hub airports for New Tokyo International Airport and Kansai International Airport, and present them as the gateways to Japan and eastern Asia.

2) Form a directional gateway network incorporating Chitose, Nagoya and Fukuoka, the major airport of the cities following the two largest metropolitan areas.

3) Form a regional network focusing on other main airports.

In order to execute these plans, massive funds are necessary for three large projects: the extension of Tokyo International Airport(Haneda airport), second phase construction of New Tokyo International Airport and the construction of Kansai International Airport. Financial problems became important due to the fact that airport rental fees are already high by international standard. Therefore, an important proposal is shown in the Aviation Council Report (1994). This report states that airport rental fees should not be raised above the present level, and that it is necessary to secure financial resource for airport improvement including the expansion of general funds. Writing of the need to increase investments in general funds seems characteristic if one reviews the processes of former transportation policies in Japan, Additionally, the report said that international hub airports are international public property and improvements should be the responsibility of Japan as a member of the international community. These ideas are quite advanced for Japan's transportation policies.

(2) Division and improvement authorities

The central and local governments control airport improvements directly. However, a public corporation makes improvements to the New Tokyo International Airport and a third-sector joint-stock company makes improvements to the Kansai International Airport. Central or local governments provide basic facilities such as runways but the terminal buildings are operated and maintained by private enterprises.

The division of airports are as follows:

1) First-class airports (Tokyo International Airport (Haneda), Osaka International Airport (Itami), New Tokyo International Airport (Narita) and Kansai International Airport)

Basically, the central government constructs and manages the airports which international airlines use and previously all expenditures were the burden of the central government. However, special legislation has been introduced gradually, and at the newly established Kansai International Airport, private funds are also used.

2) Second-class airports (26 airports such as New Chitose Airport, Sendai Airport and Fukuoka Airport)

The central government constructs and manages these airports which are used mainly by domestic airlines. If the Ministry of Transport so requests, local governments may construct and manage them.

3) Third-class airports (51 airports)

These airports are classified as necessary to secure local aviation. Local governments construct and manage them.

4) Airports shared by the Japanese Defense Agency and the United States military (6 airports such as Komatsu Airport and Miho Airport)

(3) Financial resources

A specific funds system has been implemented. Revenues from aircraft fuel tax are placed in specific funds for the central government, and a portion of aircraft fuel tax is transferred to specific funds in local governments. The central government is establishing a special account for airport improvement financed by airport rental fees such as aircraft landing fees, aircraft fuel tax, treasury investments and loans. Grant subsidies are provided from this account to airports constructed and managed by the central and local governments and invested in the New Tokyo International Airport Authority and Kansai International Airport Campany.

Local governments improve airports using aircraft fuel transferred tax, subsidies from the central government, airport rental fees and general funds.

Financial resources of the New Tokyo International Airport Authority, sources include investment from the central government, debt such as public corporation bonds and self-created funds.

The financial resources for improvements by the Kansai International Airport Company come from investments from the central and local governments, the private sector and debt.

The burden rates are set according to the Airport Improvement Act (1956). As the central government settles the Airport Improvement Plan, airports are not improved under individual income and expenditure computations, with the exception of the New Tokyo International Airport and Kansai International

Airport. The financial resources for airport improvement are from a nationwide pool system.

In this chapter, the transportation policies and systems of five developed countries have been described, and differences are recognized. The following chapters will investigate how the policies and systems influence the actual investment amounts and the shares of the actual contributors.

CHAPTER 4

A METHODOLOGY FOR COMPARING FINANCIAL RESOURCES

In this chapter, the notion of "actual" contributors is developed to allow for cross-comparison between countries having different systems and theories of finance. In the following chapters this concept is applied using statistical data from five countries to calculate the amounts invested in the improvements of transportation facilities and the contributors' shares.

4.1 PROBLEMS IN INTERNATIONAL COMPARISON OF FINANCIAL RESOURCES

4.1.1 Statistical problems

Two sets of problems arise when trying to compare financial resources across countries or across modes, namely, technical problems of incompatible statistics and methodological problems resulting from differences in countries' financial systems. The following provides a concrete explanation of the former.

(1) There is a general lack of statistical data on transportation investment and on the composition of financial resources, yet in order to make a meaningful comparison, there must be a good understanding of related investments. In this study financial resources are calculated in a comprehensive manner using many statistics, but where data is not available this is noted.

(2) Statistics are recorded for specific purposes and the content and means of collection vary according to that purpose and from country to country. An attempt has been made to ensure that the data is co-ordinated and the contents unified. Any points of ambiguity are reported in the notes. For example, interest on previous years' debt may or may not be included in the total investment given, but for our purpose we are only interested in new debt. As another example, the cost of rolling stock is generally included in railway improvement

investment data but the costs of vehicles and aircraft are not seen as part of road and airport improvement, so we deduct rolling stock from railway investment.

(3) Since the means of collecting statistics depends on their intended use, the values of even identically named data may sometimes vary and this could be a trap for the unwary, so we investigate the basis on which data is collected and make any necessary adjustments to ensure compatibility.

4.1.2 Methodological problems

Where different financial systems exist it is inappropriate to compare such conventional indices as "national expenditures" or "local expenditures". The following points should be borne in mind:

(1) Financial resources for transport improvements generally include special taxes. When such financial resources are utilized for national and local expenditures, they cannot be treated in the same way as general taxes such as income and residential taxes.

(2) The level of financial resource received as the result of user burden is different in each country; the purposes of such resources may be for a particular transport user or may be paid directly into general funds.

(3) The burden imposed on one set of users may subsidize another set, e.g. in Germany petroleum tax, a burden on car users, is used to provide financial resources for the improvement of public transport.

(4) Financial resources are sometimes derived directly from beneficiaries as in the French "Versement de Transport." Private investment may also be involved, as it is in the construction of the Kansai International Airport in Japan.

(5) There are many cases in which debt is utilized as a financial resource. Even with national government funding, the contributors will be different depending on whether the capital is refunded with general or with specific funds borne by the users. That is, the actual contributors to financial resources are those people burdened with the reimbursement of debt and they are not always the same people as those who lend the funds.

All these problems mean that conventional methods of comparison are inadequate. What is needed is a methodology that can allow comparisons from a unified viewpoint which we will discuss next.

4.2 THE BASIC CONCEPT OF CONTRIBUTOR CLASSIFICATION

4.2.1 Classification of financial resources for improvement

As shown in Table 4.1, financial resources can be divided into two main types: public funds procured by national and local governments and owner/operator's funds procured by the owner/operators themselves.

Public funds can be classified as general funds, specific funds, or debt. The owner/operator's funds can be classified as internal and self-generated funds or as external funds, such as subsidies, etc. The actual contributors of the financial resources are shown in Table 4.1.

(1) National and local tax payers

National tax payers are burdened with the general funds of the national government via income tax, corporate tax, etc. Local tax payers are those people burdened with the general funds of local governments via residence tax, fixed assets tax, etc.

(2) Users

Users are those people paying fees for using transport and taxes such as fuel tax. Thus users can be further classified as: i) users burdened with taxes collected for general funds; ii) users burdened with taxes collected for specific funds; and iii) users burdened with usage fees. In Figure 4.1 these are designated as Users (General taxes), Users (Specific taxes), and Users (Fees).

(3) Indirect beneficiaries

Indirect beneficiaries are those people who pay a special burden, not for using the transportation systems, but for external benefits such as increased property values resulting from transportation improvement.

4.2.2 Classification of contributors to debt

As shown in Figure 4.1, contributors to debt should be classified as those who pay the cost of reimbursement. There are two ways of looking at this:

(1) the amount to be invested in a given fiscal year

(2) the amount to be paid back in the fiscal year.

The amount actually invested and the amount paid back within a fiscal year are generally different. Current reimbursement is for past debt but we are

Table 4.1. Classification of financial resources.

Funds	Financial Resources	Contributors
Public funds	General funds	Payers of general funds are divided into "Payers of National Taxes" (in local government "Payers of Local Taxes"), who pay taxes as general funds, whether or not they use transportation systems, and "Users", who pay taxes as general funds when they use transportation systems.
	Specific funds	Payers of specific funds are divided into "Payers of National Taxes" (in local government "Payers of Local Taxes"), who pay taxes as specific funds whether or not they use transportation systems, "Users", who pay taxes as specific funds when they use transportation systems, and "Indirect Beneficiaries".
	Debt	Payers of reimbursement funds may vary due to the way of reimbursement.
Owners / Operators' funds	Inner funds	Inner funds are the self funds of owners / operators provided by fees and fares which users bear. Therefore, payers of inner funds are "Users".
	External funds	External funds are divided into debt refunded in later years, subsidies granted from central and local governments, investments from central and local governments, private enterprises and so on. Payers of debt become "Users in Later Years" when debt is refunded in charges and fare income from users. As for investments from central and local governments, like subsidies, payers are classified by their financial resources. Also, investments by private enterprises become "Indirect Beneficiaries" as the enterprises are blessed with benefit by improvement of the transportation systems

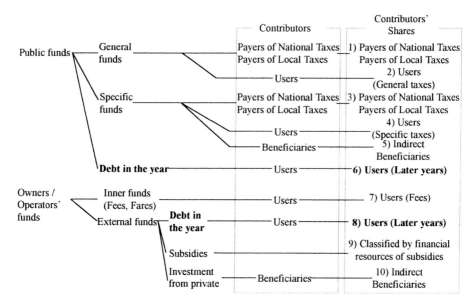

Figure 4.1. Classification of financial resources.

interested in the contribution to current infrastructural improvements, i.e. new debt, and the contributors are defined as the "people burdened with reimbursement in later years."

This method is applied in order to clarify the contributors to present infrastructure improvements.

The contributors to financial resources are divided into eight categories. An explanation of each category and examples of the financial resources are shown in Table 4.2. By separating the contributors in this way, the financial resources utilized for the improvement of transportation systems can be classified, and numerical values can be calculated for comparison among countries and transport modes.

4.3 DETERMINATION OF CONTRIBUTORS TO FINANCIAL RESOURCES

In this section, the financial resources of each country are classified by contributor category.

Table 4.2. Contributors.

Category	Financial Resource Burden	Example of Financial Resources	Classification in Fig. 4.1
Payers of National Taxes	Financial resources which citizens widely bear whether or not they use the transportation systems.	Income tax, Corporate tax, Inheritance tax, Consumption tax, Liquor tax (Japan)	1),3) Payers of National Taxes in Fig. 4.1
Payers of Local Taxes	Financial resources which residents widely bear whether or not they use the transportation systems.	Prefectural residence tax, Business tax, Fixed assets tax (Japan)	1),3) Payers of Local Taxes in Fig. 4.1
Users (General taxes)	Financial resources which users bear as general funds.	Motor vehicle tax, Light motor vehicle tax (Japan)	2) Users (General taxes) in Fig. 4.1
Users (Specific taxes, same mode)	Financial resources which users of the same transportation mode bear as specific funds.	Automobile acquisition tax, Gasoline tax (Japan)	4) Users (Specific taxes) in Fig. 4.1
Users (Specific taxes, other modes)	Financial resources which users of the other transportation modes bear as specific funds.	Petrolatum tax (Germany)	4) Users (Specific taxes) in Fig. 4.1
Users (Fees)	Financial resources which users bear as fees and fares.	Toll road fees, Railroad fares	7) Users (Fees) in Fig. 4.1
Indirect Beneficiaries	Financial resources which indirect beneficiaries bear for the purpose of improvement of transportation systems.	Private investment (Japan : Kansai International Airport), "Versement de Transport" (France)	5),10) Indirect Beneficiaries in Fig. 4.1
Users (Later years)	Financial resources which users in later years bear as fees and fares to refund debt.	Toll road fees, Railroad fares	6),8) Users (Later years) in Fig. 4.1

4.3.1 Classification of financial resources in Germany (Table 4.4)

1) Roadways

National taxpayers contribute to general funds of the federation, and local taxpayers contribute to the general funds of the state and local governments. Petroleum tax is a federal tax and automobile tax is a state tax. The portion of these taxes specified for the construction of roads corresponds to Users (Specific taxes). The portion of the taxes allocated to general funds corresponds to Users (General taxes). The public bonds issued by public works companies (ÖFFA) correspond to Users (Later years).

2) Railways

Both general funds and petroleum tax are a source of the subsidy from the federal government. This corresponds to Users (Specific taxes) because it is allocated to specific funds for subsidizing public transport.

Debt corresponds to Users (Later years) for reimbursement in the future. Fare income corresponds to Users (Fees) as it is substantively allocated to railway construction.

3) Airports

The financial resources for airport improvement from federal and local governments' general funds are classified as Payers of National Taxes and Payers of Local Taxes. Airport usage fees correspond to Users (Fees) and debt incurred corresponds to Users (Later years).

4.3.2 Classification of financial resources in France (Table 4.5)

1) Roadways

Because there is no specific funds system, all roadways other than toll roads are constructed using general funds. Taxes related to automobiles, such as registration tax and automobile tax, are also utilized for road construction, but they are general funds and therefore correspond to Users (General taxes). The other funds allocated for construction are the general funds of the central and local governments, and these correspond to Payers of National Taxes and Payers of Local Taxes. In addition, financial resources utilized by mixed-economy companies to construct toll roads correspond to the categories of Users (Fees) and Users (Later years).

2) Railways

There are no tax burdens on railway users for construction. The "Versement de Transport" is a local tax collected from enterprises that receive indirect

Table 4.3. Authorities for improvements of transportation systems.

	Roadways	Railways	Airports
Germany	**Federal government** **State and local governments**	**German Federal Railway** **_(Deutsche Bundesbahn)_** **East German National** **Railway _(Deutsche_** **_Reichsbahn)_** German Railway Corporation (after 1994.1) _(Deutsche Bahn Aktien_ _Gesellshaft)_ Private companies State and local government cf. U-Bahn	Federal government (radio facilities, weather facilities, aerial control) State and local governments (local airports) Corporations (international airports) Incorporated companies **(München International** **Airport)**
France	**Central government** **Local governments** **Mixing economy companies** **_(Société d'Economie Mixte)_** **Private approval company** **_(Société Privé_** **_Concessionnaire)_** **Special companies**	**French National Railway** **_(Socoété Nationale des_** **_Chemins de Fer Français)_** Paris transport corporation _(Régie Autonome des_ _Transport Parosiens)_ Local governments Private companies	**Central government** **Local governments** **Chamber of commerce** **Paris airport public** **_corporation (Aéroport de_** **_Paris)_** **Private companies**
United Kingdom	**Central government** **Local governments**	**British Railway Board** **London Transport** Local governments Private companies	Civil Aviation Authority **Local governments** **British Airport Authority** **Public Limited Company** Private companies
United States	**Federal government** **State and local governments** **Public corporations**	**AMTRK (National Railroad** **Passenger Corporation)** **Class I Railroad** **Transit public corporation** **(LAMTA,WMATA,MAT)** **State and local governments** Private companies	**Federal government** **State and local governments** Port/airport authorities Private companies **(Object airports of NPIAS)**
Japan	**Central government** **Local governments** **Japan Highway Public** **Corporation** **Metropolitan Expressway** **Public Corporation** **Hanshin Expressway Public** **Corporation** **Honshu-Shikoku** **Bridge Authority** **Local road public** **corporations**	**Private companies** **Special companies (JR)** **Japan Railway Construction** **Public Corporation** **Honshu-Shikoku** **Bridge Authority** **Local governments** **(Public subways)** **Teito Rapid Transit Authority** **3rd sector**	**Central government** **Local governments** **New Tokyo International** **Airport Public Corporation** **Kansai International Airport** **Corporation**

Bold bodies (underlines) are authorities included in the computation of this book.
As for other authorities, because data are not enough, we do not include in calculations.

Table 4.4. Contributors (Germany).

Payers	Roadways	Railways	Airports
Payers of National Taxes	General funds of federal government (It excludes automobile users' taxes.)	Subsidies from federal government (part of general funds)	(General funds of federal government)
Payers of Local Taxes	General funds in state and local governments (It excludes motor vehicle relation taxes.) Local bonds	Subsidies from state and local governments (part of general funds)	(General funds of state and local governments)
Users (General taxes)	Petroleum tax (part of general funds) Motor vehicle tax (part of general funds)	None	None
Users (Specific taxes)	Petroleum tax (part of specific funds) Motor vehicle tax (part of specific funds)	Subsidies from federal government (A part of specific funds) <paid by road users>	
Users (Fees)	None	(Fare income of railway owners/operators)	Airport rental fee income
Indirect Beneficiaries	None	Nothing	
Users (Later years)	Public debt (issued by ÖFFA)	Debt of railway owners/operators	Debt of airport owners/operators

benefits from construction, i.e. Indirect Beneficiaries. Other subsidies are provided by general funds from the central and local governments and therefore correspond to Payers of National Taxes and Payers of Local Taxes.

3) Airports
The subsidies from the Private Sector Aviation Special Account are covered by fees and taxes collected from airlines and passengers. These funds are classified as Users (Specific taxes) and Users (Fees), respectively. There are no definite financial resources allocated to airport construction by the central or local governments.

4.3.3 Classification of financial resources in the United Kingdom (Table 4.6)

1) Roadways
Because no specific funds have been introduced for the construction of roadways except for toll roads, all roads are constructed with general funds. Automobile-related taxes, such as automobile and fuel taxes, correspond to Users (General taxes), and other general funds of the central and local governments correspond to Payers of National Taxes and Payers of Local Taxes. The financial resources for toll roads correspond to Users (Fees) and Users (Later years).

Table 4.5. Contributors (France).

Payers	Roadways	Railways	Airports
Payers of National Taxes	General funds of central government (It excludes automobile users' taxes.)	Subsidies from central government	
Payers of Local Taxes	General funds of local governments (It excludes automobile users' taxes.)	Subsidies from local governments	
Users (General taxes)	Registration tax, Drivers license tax, Motor vehicle tax, Axle tax, Fuel tax, Corporation vehicle special tax, Load transportation tax	None	
Users (Specific taxes)	None	None	Taxes collected from airways and passengers
Users (Fees)	Fee income of the toll roads	Fare income of railway owners/operators	Airport rental fee income Fees collected from airways and passengers
Indirect Beneficiaries	None	*Versement de Transport*	None
Users (Later years)	Debt of toll road owners/operators	Debt of railway owners/operators	Debt of airport owners/operators

2) Railways

There are no tax burdens on railway users for construction. Subsidies to the London Transport (LT) and public service obligation (PSO) grants to the national railway (BR) correspond to Payers of National Taxes because they are subsidized by the general funds of the central government. The fixed asset tax (Rate) of local governments is classified as Payers of Local Taxes. Other financial resources are Users (Fees) and Users (Later years).

3) Airports

The aircraft fuel tax, which is allocated to the general funds of the central government, is a tax burden of users and corresponds to Users (General taxes). Other general funds of the central and local governments correspond to Payers of National Taxes and Payers of Local Taxes, and airport usage fees correspond to Users (Fees).

4.3.4 Classification of financial resources in the United States (Table 4.7)

1) Roadways

Automobile users are burdened with taxes, such as road-use tax, tyre tax, etc., that go into specific funds of the federal government, as well as registration

Table 4.6. Contributors (United Kingdom).

Payers	Roadways	Railways	Airports
Payers of National Taxes	General funds of central government (It excludes automobile users' taxes.)	General funds of central government Subsidies from central government PSO grant	General funds of central government Subsidies from central government
Payers of Local Taxes	General funds of local governments (It excludes automobile users' taxes.) Grant from central government Local bonds	General funds of local governments	General funds of local governments Subsidies from local governments
Users (General taxes)	Motor vehicle special tax (motorcar tax), Motor vehicle tax, Fuel tax	None	Aircraft fuel tax
Users (Specific taxes)	None	None	
Users (Fees)	Fee income of toll roads	Fare income of railway owners/operators	Airport rental fee income
Indirect Beneficiaries	None	None	
Users (Later years)	None	Debt of railway owners/operators	Debt of airport owners/operators

fees, etc., of which the funds are allocated to specific funds of the state and local governments. Therefore, these forms of financial resources are classified as User (Specific taxes). Additionally, some individuals are burdened with Tax Increment Financing (TIF), Special Assessment Districts (SAD) and Impact Fees (IF) for road construction. These forms of financial resources correspond to Indirect Beneficiaries.

2) Railways

The Intermodal Surface Transportation Efficiency Act of 1991 (ISTEA '91) established the Highway Trust Fund, from a gasoline tax which contributes to railway construction. This financial resource corresponds to Users (Specific taxes). Fare income is substantively allocated for railway construction, and is therefore classified as Users (Fees). Subsidies from the federal and local governments are provided from general funds, and therefore correspond to Payers of National Taxes and Payers of Local Taxes, respectively.

3) Airports

The federal government subsidizes airport construction with funds from the Airport and Airway Trust Fund (AATF). The capital of this fund is created by interest income from deposits and through such taxes as the ticketing or

Table 4.7. Contributors (United States).

Payers	Roadways	Railways	Airports
Payers of National Taxes	General funds of federal government (It excludes automobile users' taxes.) Subsidies from federal government	General funds of federal government Subsidies from federal government	
Payers of Local Taxes	General funds of state and local governments (It excludes automobile users' taxes.) Local bonds	General funds of state and local governments Subsidies from state and local governments	General funds of state and local governments General fund bonds
Users (General taxes)			None
Users (Specific taxes)	Motor vehicle fuel tax, Use tax, Tire tax, Trucks and trailers tax, Motor vehicle registration tax, Drivers license tax	Highway Trust Fund <paid by road users>	Airport and Airway Trust Fund (Ticketing tax, Leaving tax, Aerial freight tax, Aerial fuel tax)
Users (Fees)	Fee income of toll roads	(Fare income of railway owners/operators)	Airport rental fee income Airport operation income (Landing fee, ticketing tax, aviation freight tax, departure tax, aviation fuel tax)
Indirect Beneficiaries	Contribution from beneficiaries	Private investment	
Users (Later years)	Debt of toll roads owners/operators	Debt of railway owners/operators	Debt of airport owners /operators (revenue bonds)

departure taxes which are classed as Users (Specific taxes). Airport usage fees correspond to Users (Fees). Revenue bonds are also used by the local governments, but reimbursement comes from airport income thus corresponding to Users (Later years).

4.3.5 Classification of financial resources in Japan (Table 4.8)

1) Roadways

Automobile-related taxes, such as gasoline tax, petroleum tax, etc., provide capital for specific funds of the central and local governments, and correspond to Users (Specific taxes). Other taxes, such as automobile taxes, light-automobile taxes, etc., are local taxes allocated to general funds and are therefore Users (General taxes). Although revenues from the automobile tonnage tax go into general funds, the capital corresponds to Users (Specific taxes), because it is substantially used for road construction.

Treasury investment and loans are utilized by four public corporations associated with the improvement of toll roads. These funds are treated as debt for reimbursement through fees charged to users in later years, and thus classified as Users (Later years).

Subsidies provided from the general funds of central and local governments correspond to Payers of National Taxes and Payers of Local Taxes, respectively. Although subsidies provided from the general funds of the central government are very rare, this method is included as a possible means of roadway finance.

2) Railways

There are no tax burdens on railway users to derive for construction. The central government used part of the capital gains from the privatization of the Shinkansen (Superexpress) to establish the Railway Development Fund, a specific fund for railway construction. From this fund, subsidies are paid to construct new Shinkansen, new trunk lines, and urban railways. However, because this fund is not collected as a tax, but is covered by burden to the users of Japan Railways (JR) it corresponds to Users (Fees).

Subsidies provided by the central and local governments are derived from general funds and therefore correspond to Payers of National Taxes and Payers of Local Taxes, respectively.

Funds procured privately by the railway companies are classified as Users (Fees) and Users (Later years).

Table 4.8. Contributors (Japan).

Payers	Roadways	Railways	Airports
Payers of National Taxes	General funds of central government (It excludes automobile users' taxes.) Government bonds	Subsidies from central government	General funds of central government Subsidies from central government
Payers of Local Taxes	General funds of local governments (It excludes automobile users' taxes.) Local bonds of general accounts	Subsidies from local governments Investments from local governments	General funds of local governments Subsidies from local governments
Users (General taxes)	Motor vehicle tax, Light motor vehicle tax,	None	Aircraft fuel tax
Users (Specific taxes)	Gasoline tax, Gas oil delivery tax, Petroleum gas tax, Petroleum gas transferred tax, Motor vehicle weight tax, Motor vehicle weight transferred tax, Local road tax, Automobile acquisition tax	None	
Users (Fees)	Fee income of toll roads	Specific funds in central government (Revenues created from selling existing Shinkansen) Fare income of railway owners/operators Capital increase of railway owners	Airport rental fee income
Indirect Beneficiaries	None	None	Private investments
Users (Later years)	Debt of toll road owners/operators (It includes treasury investment and loans)	Debt of railway owners/operators	Debt of airport owners/operators (It includes treasury investment and loans)

3) Airports

The aircraft fuel tax revenues go into a specific fund of the central government, and the aircraft fuel transferred tax revenues constitute a specific fund of the local governments, thereby corresponding to Users (Specific taxes), since the burden is placed on the users. Other financial resources for the improvement of airports include investment by private companies that benefit from the use of the airports classified as Indirect Beneficiaries.

Treasury investments and loans utilized as financial resources from special accounts for airport improvement are classified as Users (Later years), since reimbursement of the debt is to be covered by fees burdened on the users in later years.

Subsidies from general funds are provided by the central and local governments, and therefore correspond to Payers of National Taxes and Payers of Local Taxes, respectively.

CHAPTER 5

THE CALCULATION OF INVESTMENT AMOUNT

5.1 METHOD OF CALCULATION

In this chapter we calculate the level of improvement investment for each mode of transport in our sample of developed countries. Special attention is given to: 1) unification of the data, and 2) confirmation of data reliability by utilizing more than one set of statistics.

The procedure applied for the purpose of calculation is as follows: 1) total amount of actual investment per year, 2) investment split into construction, and repairs and maintenance, 3) the composition of financial resources. The latter covers such categories as National Expenditures, local expenditures, owners/operators' funds and debt rather than substantive contributors. This is explained in the following chapter.

5.2 ROADWAYS

5.2.1 Roadway Investment in Germany

Statistical Data sources
World Road Statistics (International Road Federation)
Verkehr in Zahlen (Der Bundesminister für Verkehr[German Federal Ministry of Transport])
Straßenbaubericht (Der Bundesminister für Verkehr[German Federal Ministry of Transport])
West German Road Society Data (West German Road Society)
Statistisches Jahrbuch (Statistishes Budesamt [German Federal Statistics Agency])

(1) Calculation of total investment

Federal, state and local governments all have responsibility for improving roadways. The data sources used are: "World Road Statistics," published by the

Table 5.1(1). Investment amounts for roadway improvements and financial resource composition.

		Investment amount	National expenditures	Local expenditures	Owners / Operators' funds	Debt	New construction	Maintenance and repaire
Germany	Total	34,949	9,859	25,090			83.1%	16.9%
	Bundes autobahnen (Autobahn)	5,284	4,803	481				
	Bundesund LandesStraßen (Federal and state roads)	9,204	4,409	4,795				
	Kreis straßen (District roads)	1,539						
	Gemeinde straßen (City, town and village roads)	15,157	647	19,814				
	Verwaltung und Sonstiges (Administration, others)	3,765						
France	Total	71,308	11,040	45,601	2,526	12,141	76.8%	23.2%
	Central and local governments	56,641	11,040	45,601				
	Toll Roads	14,667			2,526	12,141		
United Kingdom	Total	4,795	1,958	2,837	0	0	60.8%	39.2%
Unites States	Total	75,338	15,523	48,737	2,998	8,080	69.0%	31.0%
	On State-Administered Highways	35,890	7,395	23,218	1,428	3,849		
	On Locally Administered Roads	24,348	5,017	15,751	969	2,611		
	Others	15,100	3,111	9,768	601	1,619		
Japan	Total	12,815	3,210	6,851	937	1,818	82.6%	17.4%
	General roads and so on	9,854	3,077	6,778	0	0		
	Local road public corporations and so on	226	0	49	5	172		
	4 public corporations	2,735	133	24	932	1,646		

Unit, Year	Germany	Million DM	1992
	France	Million FF	1993
	U.K.	Million £	1991
	U.S.A.	Million US$	1992
	Japan	Billion Yen	1992

Table 5.1(2). Investment amounts for roadway improvements and financial resource composition (GDP ratio).

	Investment amount	National expenditures	Local expenditures	Owners / Operators' funds	Debt
Germany	1.251%	0.353%	0.898%	0.000%	0.000%
France	1.005%	0.156%	0.643%	0.036%	0.171%
United Kingdom	0.835%	0.341%	0.494%	0.000%	0.000%
United States	1.251%	0.258%	0.810%	0.050%	0.134%
Japan	2.763%	0.692%	1.477%	0.202%	0.392%

International Road Federation, and "Verkehr in Zahlen," published by Der Bundesminister für Verkehr (German Federal Ministry of Transport). Of the two, "Verkehr in Zahlen" gives the most detail. For the total amount invested, the Nettoausgaben des Bundes der Lander und der Gemeinden für das Straßenwesen (Net Road Expenses of the Federation, State and Local Governments) from "Verkehr in Zahlen," provides the most accurate value: DM 34,949 million in 1992.

Transport police (Verkehrpolizei) are not an investment in road improvement, so this amount should be deducted from the total when attempting to make comparisons with other countries.

The content of the total amount invested in road improvement by category is: autobahn (Bundesautobahnen) DM 5,284 million; federal and state roads (Bundesundlandstraßen) DM 9,204 million; district roads (Kreisstraßen) DM 1,539 million; towns and villages roads (Gemeindestraßen) DM 15,157 million; and other management operations (Verwaltung und Sonstiges) DM 3,765 million. Maintenance and repairs are also included in these figures. The ratio of maintenance and repairs for federal long-range roads is 16.9%.

(2) Calculation of financial resource composition

The National Expenditures are DM 9,859 million as mentioned in the section of Verkehrsausgaben Bundesfernstraßen (Transportation Expense of the Federal Government, Federal Long-Range Roads) in "Verkehr in Zahlen." This value is in agreement with that in "Straßenbaubericht," which is published by Der Bundesminister für Verkehr (German Federal Ministry of Transport).

Since there are no financial resources classified as owners/operators' funds in the German system, the remaining amount is local expenditures.

5.2.2 Roadway Investment in France

Statistical Data Sources
World Road Statistics (International Road Federation)
Mémento de la Route (Ministére de l'Equipment Direction des Routes des Transports et du Tourisme Direction des Routes)
Construction Situation of Inter-City Expressways in Every Country (Expressways and Automobiles)
Annuaire Statistique de la France (Ministére de l'Economie)

(1) Calculation of total investment

Central and local governments are the authorities for the improvement of roadways, and the mixed-economy companies (SEMs) are the authorities for toll roads.

(2) Calculation of financial resource composition

Data can be found in "Mémento de la Route," published by the Ministére de l'Equipment Direction des Routes, and "World Road Statistics," published by the International Road Federation. The amount invested by central government as well as the amount invested in toll road construction is given in the former as well as in "Construction Situation of Inter-City Expressways in Every Country" (Expressways and Automobiles, 1993), and the amounts invested by all authorities are in the latter. Central government investment listed in the Mémento de la Route and government expenses listed in "Construction Situation of Inter-City Expressways in Every Country" are identical, namely FF 11,040 million.

The investments by the local governments listed in "World Road Statistics" under Local Governments and Municipalities totals FF 45,601 million.

Statistics for the construction of toll roads by mixed-economy companies (SEMs) are given in "Mémento de la Route," and "World Road Statistics." Since the French government does not use debt to construct toll expressways the value of "Mémento de la Route", FF 12,141 million, is the invested amount of debt by mixed-economy companies (SEMs). The addition of owners/operators' funds is given as FF 2,526 million from Income Applied to Maintenance Expense.

Calculated investment includes maintenance and repair expenses (Sous total entretien) for both central government and mixed-economy companies (SEMs). The rate of new construction is 76.8%, and the rate of maintenance and repair is 23.2%.

5.2.3 Roadway Investment in the United Kingdom

Statistical Data Sources
World Road Statistics (International Road Federation)
Transport Statistics Great Britain (Department of Transport)
Transport Statistics Report of Road Traffic Statistics Great Britain
(Department of Transport)
Annual Abstract of Statistics (HMSO)

(1) Calculation of total investment

Central and local governments are the improvement authorities for road-ways. Data can be found in "World Road Statistics" (International Road Federation), "Transport Statistics Great Britain" (Department of Transport), "Transport Statistics Report Road Traffic Statistics Great Britain" (Department of Transport) and the "Annual Abstract of Statistics" (HMSO). "World Road Statistics," published by the International Road Federation, lists the total National Expenditures in investment as £3,375 million, and the local expenditures of local governments and municipalities as £4,890 million.

The rate of new construction is 60.8%, and the rate of maintenance and repair expense is 39.2%.

5.2.4 Roadway Investment in the United States

Statistical Data Sources
World Road Statistics (International Road Federation)
Highway Statistics (Federal Highway Administration)
Statistical Abstract of the United States (Department of Commerce, Bureau of Census)

(1) Calculation of total investment

The road improvement authorities are the federal, state and local governments, and public corporations that improve toll roads. The investment amounts for road improvement can be found in the "World Road Statistics" (International Road Federation), "Highway Statistics" (Federal Highway Administration) and the "Statistical Abstract of the United States" (Department of Commerce, Bureau of Census).

The amount invested in roadways is listed under the Grand Total Disbursement. This includes bond retirements and interest on debt which must be subtracted to give a total investment of U.S.$ 75,338 million.

Financial resources are categorized as National Expenditures, local expenditures, owners/operators' funds and debt. Using the values of revenue in "Funding for Highways and Disposition of Highway-User Revenue, the composition ratios can be calculated. The composition of the financial resources are calculated by multiplying the composition ratios by the investment amounts. The ratios calculated are 20.6% for National Expenditures, 64.7% for local expenditures, 4.0% for owners/operators' funds and 10.7% for debt.

5.2.5 Roadway Investment in Japan

Statistical Data Sources
Road Statistics Yearbook (National Road Users Conference)
Japan Highway Public Corporation Yearbook
(Japan Highway Public Corporation)
Japan Highway Public Corporation Data
(Japan Highway Public Corporation)
Metropolitan Expressway Public Corporation Yearbook
(Metropolitan Expressway Public Corporation)
Hanshin Expressway Public Corporation Yearbook
(Hanshin Expressway Public Corporation)
Honshu-Shikoku Bridge Authority Data
(Honshu-Shikoku Bridge Authority)
Reference Calculation Data for Local Taxes (Ministry of Home Affairs)
Financial Statistics (Ministry of Finance)

Roadways are improved by central and local governments, and by public corporations such as the Japan Highway Public Corporation, specified urban expressway corporations and local road public corporations. Central and local governments improve general roads for which no user fees are collected. Toll roads are improved by either the four public corporations, specific urban expressway public corporations, local road public corporations or local governments.

(1) Calculation

1) General road improvement

The most detailed data for the amount of investment in road improvement is in the "Road Statistic Yearbook," published annually by the National Road Users Conference, as well as the data provided by the four public corporations related to roads. The investment amounts mentioned in the "Road Statistic Yearbook" is divided into the following categories based on responsibility for expenses: Central government, Prefectures, Towns and villages, and Others. The expense share from beneficiaries or responsible persons except for the local governments (Nihon Telegraph and Telephone Corporation, gas/water/electric power utilities, railway companies, etc.) are appropriated to "Others." Such expenses are not those allotted to road improvement as transportation systems, therefore they should not be included. Investment calculated in this study is ¥9,854 billion.

2) Toll road improvement by specific urban expressway public corporations

The investment amount for toll road improvement by specific urban expressway public corporations is provided in the "Road Statistic Yearbook." The items listed are classified as Construction expenses and Maintenance and repair expenses, and the sum of each item added together gives the total investment amount.

3) Toll road improvement by local road public corporations and local governments

The investment amounts for toll road improvement by local road public corporations and local governments are provided in the "Road Statistic Yearbook," and are calculated as mentioned in 2). The total amount of investment derived by adding 2) and 3) is ¥226 billion.

4) Toll road improvement by four public corporations related to roads

The investment amounts for toll road construction by the four road-related public corporations are in the yearbooks published by each of the corporations. The total amount of investment is ¥2,734 billion.

Expense covered by beneficiaries and responsible persons, excluding the local governments, is included in the maintenance and repair expenses of general road improvement, so the investment amounts are divided utilizing the appropriate ratio of construction to maintenance, and the maintenance and repair expenses are calculated. The resulting share of investment spent on new construction is 82.6% and that of maintenance and repair is 17.4%.

(2) Calculation of financial resource composition

1) General road improvement
    National and local expenditures are utilized for general road improvement. The figures listed under central government in the "Road Statistic Yearbook" are National Expenditures and those listed under Prefectures, Towns and villages are local expenditures.

2) Toll road improvement by specific urban expressway public corporations
    The financial resources available to specific urban expressway public corporations include local expenditures, owners/operators' funds and debt. Since maintenance and repair expenses are covered by revenues from fees charged, investment from prefectures and cities in "Road Statistic Yearbook" becomes the local expenditures, and maintenance and repair expenses become owners/operators' funds. The remaining amount is debt.

3) Toll road improvement by local road public corporations and local governments
    Local expenditures and debt are utilized for toll road improvement by local road-related public corporations and local governments. National Expenditures are not utilized. The amount of local expenditures and of debt can be calculated from the "Road Statistic Yearbook."

4) Toll road improvement by the four public corporations related to roads
    The four road-related public corporations utilize investments and subsidies from the central government, from local governments and from local government bonds. Owners/operators' funds come from business income and debt, such as government loans and guaranteed government bonds. Although the total amounts are taken from the yearbooks of each public corporation, expenses which are not directly related to road construction, such as operating expenses or non-business expenses are also included, so have to be subtracted.

    The amounts invested in road improvement and the composition of financial resources are shown in Table 5.1(1).

5.3 RAILWAYS

    There are many authorities in every country that improve railways. It is very difficult to calculate the amount of investment and the composition of financial resources of all of the authorities because the statistical data fails to cover all of them. However, in this section, an attempt is made to cover the

Table 5.2(1). Investment amounts for railway improvements and financial resource composition.

		Investment amount	National expenditures	Local expenditures	Owners / Operators' funds	Debt	New construction	Maintenance and repaire
Germany	Total	13,904	10,204	0	2,468	1,231	58.72%	41.28%
	Deutsche Bahn (DB)	7,159	3,564	0	2,396	1,199	47.78%	52.22%
	Deutsche Reichsbahn (DR)	6,745	6,640	0	72	33	70.35%	29.65%
France	Total	20,521	11,565	1,092	4,489	3,374	80.93%	19.07%
	SNCF	16,812	10,871	551	3,476	1,914	79.33%	20.67%
	RATP	3,709	694	542	1,013	1,460	88.20%	11.80%
United Kingdom	Total	2,344	975	0	1,207	162	51.36%	48.64%
	British Railway Board	1,717	595	0	960	162	42.88%	57.12%
	London Transport	627	380	0	247	0	74.59%	25.41%
United States	Total	9,651	1,771	1,430	6,451	0	61.24%	38.76%
	Class I Railroads	5,738	5	0	5,733	0	48.71%	51.29%
	Amtrak	512	480	0	31	0	28.86%	71.14%
	Transit	3,402	1,286	1,430	686	0	87.23%	12.77%
Japan	Total	1,726	195	167	689	675	79.42%	20.58%
	Private railways	978	28	12	560	379	80.28%	19.72%
	Public subways	471	50	129	80	212	66.09%	33.91%
	Japan Railway Construction Corporation	276	117	26	49	84	99.10%	0.90%
	Honshu-Shikoku Bridge Authority	0	0	0	0	0	100.00%	0.00%

Unit, Year			
	Germany	Million DM	1993
	France	Million FF	1993
	U.K.	Million £	1993
	U.S.A.	Million US$	1993
	Japan	Billion Yen	1992

Table 5.2(2). Investment amounts for railway improvements and financial resource composition (GDP ratio).

	Investment amount	National expenditures	Local expenditures	Owners / Operators' funds	Debt
Germany	0.491%	0.360%	0.000%	0.087%	0.043%
France	0.289%	0.163%	0.015%	0.063%	0.048%
United Kingdom	0.374%	0.156%	0.000%	0.193%	0.026%
United States	0.152%	0.028%	0.023%	0.102%	0.000%
Japan	0.372%	0.042%	0.036%	0.149%	0.146%

large-scale authorities, namely, in Germany—German Federal Railway (DB) and the former East German National Railway (DR); in France—French National Railway (SNCF) and Paris Transport Corporation (RATP); in the United Kingdom—British Railways Board (BR) and London Transport (LT); in the United States—Class I Railroads, Amtrak (the sole inter-city passenger railway in the U.S.) and public passenger transportation (Transit), which local governments manage; and in Japan—All railway authorities. The amount of investment and the composition of financial resources for railway improvement are calculated utilizing the methods explained below.

(1) Methods for calculating total investment

The items included under investment vary from source to source so an attempt has been made to unify the definition of total investment, i.e. capital investment consisting of capital expenditures and maintenance and repair expenses. Rolling stock and operational expenses are excluded.

Maintenance and repair, which are generally appropriated to operating expenses (profit and loss statement), are included because these are substantive investments in the railway infrastructure.

(2) Methods for calculating financial resource composition

In most cases, the composition of financial resources for railway investment are not shown separately in the data but are combined with operations, so the composition of the financial resources were calculated as follows.

1) The total revenue of railway constructors is broken down into six categories covering both investment and operating expenses: i) capital grants by central government, ii) capital grants by local governments, iii) operating grants by central government, iv) operating grants by local governments, v) operation income, and vi) debt.

2) Since capital grants from central and local governments are allotted for investment, these grants correspond to National Expenditures and local expenditures, respectively.

3) The amounts remaining, excluding the capital grants from total amounts of investment, are outlays from other income. Therefore, they are divided using the rates of operating income and debt in the total revenue.

Incidentally, when rolling stock expenditures are included in the amount of investment, they are deducted after calculating the composition of the investment amount.

5.3.1 Railway investment in Germany

Statistical Data Sources
Data of Foreign Railway Investigations—No. 36 (JR Group)
Present Situation of Foreign Railways (JR Group)
Verkehr in Zahlen (Der Bundesminister für Verkehr)
World Railway Investment (IRJ)
Company Report (Deutsche Bahn)
Bericht über das Geschäftsjahr (Deutsche Bahn)

(1) Calculation of total investment

Data for railway improvement by the German Federal Railway (Deutsche Bundesbahn: DB) and the former East German National Railways (Deutsche Reichsbahn: DR) are provided in the "Bericht über das Geschäftsjahr: Deutsche Bahn," "Moving with Times: Deutsche Bahn" and "Facts and Figures: Deutsche Bahn." The "Data of Foreign Railway Investigations" based on these publications.

When making calculations using data from the "Present Situation of Foreign Railways," published by the JR group, "Company Report," published by Deutsche Bahn, and the above sources, the investment amount in 1993, excluding rolling stock expenditures, was DM 13,904 million.

(2) Calculation of financial resource composition

The composition of financial resources is calculated, first with, then without rolling stock. The composition of financial resources is mentioned in detail in the "Bericht über das Geschäftsjahr," "Data of Foreign Railway Investigations" and "Present Situation of Foreign Railways."

5.3.2 Railway investment in France

Statistical Data Sources
Data of Foreign Railway Investigations—No. 38 (JR Group)
Present Situation of Foreign Railways (JR Group)
Present Situation and Future of the French National Railway
World Railway Investment (IRJ)
Mémento de Statistiques Exercice (SNCF)
French Railway Statistics Summary (SNCF)
French National Railway Annual Report (SNCF)

(1) Calculation of total investment

Investment and financial resource composition was calculated for French National Railway (SNCF) and Paris Transport Corporation (RATP).

1) French National Railway (SNCF)
Detailed data is provided in the "Present Situation of Foreign Railways," "Data of Foreign Railway Investigations" and "Mémento de Statistiques Exercice." Some of the measures of investment include the "Programme spécial banlieu-frais généraux unclus (Special Paris Suburb Plan)" so adjustment is needed to account for this. The total amount of investment is FF 20,521 million.

2) Paris Transport Corporation (RATP)
The detailed data on RATP's investment levels is given in "Urban Transportation Information Annual Report; Paris Transport Corporation" but includes business other than railways (e.g., busing) which has to be subtracted from the total.

The total for maintenance and repair can be found in the section under Operating Improvement in the section Capital Investment Accounts of the report.

(2) Calculation of financial resource composition

1) French National Railway (SNCF)
Detailed information on the composition of financial resources of SNCF is provided in the "French National Railway Annual Report," "French Railway Statistics Summary," "Data of Foreign Railway Investigations" and "Present Situation of Foreign Railways." The ratios of each financial resource item for investment, including rolling stock expenditures, are first calculated, and then the values for rolling stock expenditures are removed.

2) Paris Transport Corporation (RATP)
The composition of financial resources of the RATP can be calculated from the figures in the section Capital Investment Account provided in the "Urban Railway Transport Information;" operating expenses are not included in the data.

5.3.3 Railway investment in the United Kingdom

Statistical Data Sources
Overseas Urban Transportation Information (London Transport Annual Report) (Teito Rapid Transit Authority)
Data of Foreign Railway Investigation—No. 40 (JR Group)
Present Situation of Foreign Railways (JR Group)
World Railway Investment (IRJ)
Great Britain Transport Statistics (Department of Transport)
London Transport Annual Report (London Transport)
Annual Report and Accounts (British Railways Board)

(1) Calculation of total investment

The investment amount and financial resource composition were calculated for British Railways Board (BR) and London Transport (LT).

1) British Railways Board (BR)
BR data is given in "Data of Foreign Railways Investigation," "Great Britain Transport Statistics" and "Present Situation for Foreign Railways." The most detailed information is provided in the "Annual Report and Accounts," published by the British Railways Board. The amount of investment was £1,717 million.

2) London Transport (LT)

LT investment was calculated using "Great Britain Transport Statistics" and the "Overseas Urban Transportation Information," prepared by the Teito Rapid Transit Authority and based on the "London Transport Annual Report." The total investment amount was £627 million.

(2) Calculation of financial resource composition

1) British Railways Board (BR)

Detailed information provided in "Data of Foreign Railways Investigation," and the composition is calculated utilizing the numerical values listed therein. The appropriation of debt is listed in the "Annual Report and Accounts" (British Railroads Board).

2) London Transport (LT)

The composition of financial resources is calculated by subtracting the rolling stock expenditures after calculating the ratio of financial resources to investment. Detailed data is provided in the London Transport Annual Report.

5.3.4 Railway investment in the United States

Statistical Data Sources
Data of Foreign Railway Investigations—No. 39 (JR Group)
Present Situation of Foreign Railways (JR Group)
Statistical Abstract of the United States (Department of Commerce, Bureau of Census)
World Railway Investment (IRJ)
Railroads Facts (Association of American Railroads)
Analysis of Class I Railroads (Association of American Railroads)
Annual Report (National Railroad Passenger Corporation)
Data Tables of the National Transit Database Section 15 Report Year (Federal Transit Administration)

(1) Calculation of total investment

The amount invested and financial resource composition were calculated for Class I Railroads and Amtrak, and Public transportation (Transit).

1) Class I Railroads and Amtrak

The data for investment amount of Class I Railroads is provided in "Present Situation of Foreign Railways," "Statistical Abstract of the United

States," "Railroad Facts" and "Analysis of Class I Railroads." Amtrak data is provided in "Railroad Facts," "Data of Foreign Railway Investigations," and "Annual Report" of the National Railroad Passenger Corporation (Amtrak). The investment amounts were U.S.$ 5,977 million for Class I Railroads and U.S.$ 512 million of Amtrak.

2) Public transportation (Transit)
The data source is "Data Tables of the National Transit Database," published by the Federal Transit Administration and the relevant categories are Automated Guideway, Cable Car, Commuter Rail, Heavy Rail, Light Rail and Monorail. The amount of investment was U.S.$ 3,402 million.

(2) Calculation of financial resource composition

1) Class I Railroads, Amtrak
   Details are provided in "Analysis of Class I Railroads." Grants from the federal and local governments, listed as Transfers from Government Authorities in the income statement of "Analysis of Class I Railroads," is regarded as operating grants of the federal government. However, the bulk of government transfers are the contract payments from various governments to Class I railroads for providing passenger services. Government Grants listed in "Data of Foreign Railway Investigations" are regarded as National Expenditures, and the remaining amount is owners/operators' funds.

2) Public transportation (Transit)
   The composition of the financial resources was calculated from "Data Tables of the National Transit Database Section," which in turn were used to calculate investment amounts.

5.3.5 Railway investment in Japan

Statistical Data Sources
World Railway Investment (IRJ)
Investigation Report of Capital Investment Trend by Enterprises Related to Transportation (Ministry of Transport)
Subway (Separate volume, Japan Subway Society)
Financial Tables (Japan Railway Construction Public Corporation)
Data from the Honshu-Shikoku Bridge Authority
Comprehensive Bibliography of Subsidies (Financial Investigation Meeting)
Japan National Railways Inspection Report (Japan National Railways)
Transport Almanac (Transport Newspaper)

Investment amounts and financial resource composition was calculated for each of the following authorities: i) private railway companies including Japan Railways (JR), ii) Teito Rapid Transit Authority and local governments (public subways), iii) Japan Railway Construction Public Corporation, and iv) Honshu-Shikoku Bridge Authority. Also included is the calculation for the former Japan National Railway (JNR).

(1) Calculation of total investment

1) Private railway companies
The details for the investment of private railway companies are provided in "Investigation Report of Capital Investment Trend by Enterprises Related to Transportation." The capital investment amounts in this data are listed as Expense Based and Construction Based. Investment was calculated using the expense-based values. However, the items listed only as construction based were converted to expense based.

2) Teito Rapid Transit Authority and local governments (public subways)
The investment amounts are provided for each improvement authority in "Subway (separate volume)," which gives data for the Teito Rapid Transit Authority and public subways. The investment amounts for rolling stock are included in these values, so they were subtracted from the total.

3) Japan Railway Construction Public Corporation
The details for the investment amounts of Japan Railway Construction Public Corporation are provided in each annual edition of the "Japan Railway Construction Public Corporation Financial Tables."

4) Honshu-Shikoku Bridge Authority
The details for the investment amounts of the Honshu-Shikoku Bridge Authority are provided in "Honshu-Shikoku Bridge Authority Data." The Construction Expenditures and Investigation Expenditures listed in this data are utilized as the investment amounts.

Since Japan Railway Construction Public Corporation and Honshu-Shikoku Bridge Authority do not manage railway operation, there are no expenditures for rolling stock.

(2) Calculation of financial resource composition

1) Private railway companies

The information regarding the procurement of capital investment funds for private railway companies is provided in each edition of "Investigation Report for Capital Investment Trend by Enterprises Related to Transportation." However, the income created from fees charged by companies and the subsidies from central and local governments are lumped together and appropriated as Inner Funds. Therefore, the values of subsidies to private railway companies provided in the "Comprehensive Bibliography of Subsidies" are also utilized.

2) Teito Rapid Transit Authority and local governments (public subways)

The information regarding financial resources is provided in separate volumes of "Subway" for each improvement authority of the Teito Rapid Transit Authority and public subways.

3) Japan Railways Construction Public Corporation

The information about financial resources for the Japan Railway Construction Public Corporation is provided in "Japan Railway Construction Public Corporation Financial Tables."

4) Honshu-Shikoku Bridge Authority

The information about financial resources for the Honshu-Shikoku Bridge Authority is provided in "Honshu-Shikoku Bridge Authority Data." The system provides no financial resources from local governments.

(3) Former Japan National Railway (JNR)

Detailed information on investment and financial resource composition for the former Japan National Railway are provided in the "Japan National Railways Inspection Report" and the "Transport Almanac." Subsidies from local governments were not granted to the former JNR, and there were no capital or operating grants from local governments.

The amounts invested in railway improvement and the composition of the financial resources are shown in Table 5.2(1).

5.4 AIRPORTS

There is little comprehensive data on the investment because of the variety of improvement authorities and constructors (National government, local governments, airport authorities, private companies, etc.) but little participation by central governments. As most investment occurs at major airports this data is more reliable and is used for our calculation.

5.4.1 Airport investment in Germany

Statistical Data Sources
Airport Construction in Every Country (European Volume, Airport Problem Study Session)
Construction Techniques and Operation of Major Airports Overseas (Japanese Project Industry Conference)
Verkehr in Zahlen (Der Bundesminister für Verkehr)

There is little data except for the "Verkehr in Zahlen: Der Bundesminister für Verkehr" (German Federal Ministry of Transport). Other than the new Munich airport, there are no new airport construction plans so the investment amounts and financial resource composition for the new Munich airport were calculated.

5.4.2 Airport investment in France

Statistical Data Sources
Activité des aéroports français (Direction Générale de l'Aviation
Civile-Service des Bases Aériennes Bureau de la Gestion des Aéroports)

The details of investment amounts for airport construction and financial resource composition in France are provided in "Activité des aéroports français" published by the "Ministére de l'Equipment, des Transports et du Tourisme Direction Générale de l'Aviation Civile-Service des Bases Aériennes Bureau de la Gestion des Aéroports." All of the airports including overseas airports listed in the data are utilized, and investment and financial resource composition is calculated. Subsidies from the EEC in this data were treated as capital grants from the central government.

The Capital Investment Amounts provided in the data are regarded as investments in airport construction. The Investment Amounts were divided proportionally at the rate of each financial resource item, and the composition of financial resources calculated. Maintenance and repairs are not included in the calculation because of data limitations. The investment amounts for Paris Airport Public Corporation (ADP) and the composition of financial resources was calculated in the same way.

5.4.3 Airport investment in the United Kingdom

Statistical Data Sources
Great Britain Transport Statistics: The Department of Transport
Construction Techniques and Operating Major Airports Overseas (Japanese Project Industry Conference)

The data for investment amounts for airports in the United Kingdom are provided in "Great Britain Transport Statistics: The Department of Transport." Data related to investment in facilities by British Airport Authority Public Company Limited (BAAplc) is provided in "Construction Techniques and Operating of Major Airports Overseas." The amount of capital investment by BAAplc, excluding the investment in Heathrow Express, is considered the investment amount. Maintenance and repair expenses are excluded from the investment amount.

5.4.4 Airport investment in the United States

Statistical Data Sources
National Plan of Integrated Airport Systems (NPIAS: Congressional Report, Federal Aviation Administration)
FAA Statistical Handbook of Aviation (Federal Aviation Administration)

Airports in the United States are improved mainly in accordance with the "National Plan of Integrated Airport Systems: NPIAS" (incorporates 3,285 of the 5,598 public-use airports). Therefore, a combination of the total investment amount for airports covered by the NPIAS and investments (maintenance and operating expenses such as air-traffic control and facility construction) by the Federal Aviation Administration is taken as the investment amount for airport improvement in the United States.

The data for airport expenses is included in the NPIAS congressional report "National Plan for Integrated Airport Systems 1993–1997," published by the FAA. Information on the FAA budget is provided in the "FAA Statistical Handbook."

5.4.5 Airport investment in Japan

Statistical Data Sources
Seeing Aviation by the Figures (Aerial Promotion Fund)
Data of the New Tokyo International Airport Authority
Airport Handbook (Kansai Airport Investigation Meeting)
Financial Statistics (Ministry of Finance)

Annual Statistics Report on Local Government Finance
(Local Government Financial Society)

(1) Calculation of total investment

Japan has a five-year plan for airport improvement. Airport improvement expenses are included in the five-year plan, however maintenance and operating expenses as outlays from the Airport Improvement Special Accounts as well as the maintenance expense of local airports by local governments are excluded. Investments by the New Tokyo International Airport Authority and the Kansai International Airport Corporation are only partly included in the five-year plan. The investment amounts and financial resource composition are broken up into four categories and then calculated: 1) Improvement by the five-year plan, 2) New Tokyo International Airport Authority, 3) Kansai International Airport Corporation, and 4) Other airport maintenance.

1) Improvement by the five-year plan
The data is provided in "Seeing Aviation by the Figures." (The New Tokyo International Airport Authority and Kansai International Airport Corporation are dealt with separately.)

2) New Tokyo International Airport Authority
The detailed data for investment amounts are provided by the authority. As this data also provides expenses other than airport improvement, the items in the data are classified for excluding them.

3) Kansai International Airport Corporation
The data for the Kansai International Airport Corporation are provided in the "Airport Handbook."

4) Other airport maintenance
Airport maintenance is not included in the investment amounts for improvement listed in the five-year plan calculated in 1). Therefore, the maintenance expense of airports excluding the New Tokyo International Airport and Kansai International Airport, which were calculated in 2) and 3), is appropriated in the "Maintenance and Operating Expenses" of the "Airport Improvement Special Accounts", and in "Local Airport Maintenance Expense" of the "Annual Statistics Report on Local Government Finance."

(2) Calculation of financial resource composition

The composition of financial resources is calculated in the same manner as that utilized to calculate the investment amounts (i.e., by four categories).

1) Construction by the five-year plan
The financial resources for improvement carried out under the five-year plan are covered by the general funds of local governments and the Airport Improvement Special Accounts. A more detailed explanation of the financial resources of the Airport Improvement Special Accounts is provided in "Financial Statistics."

2) New Tokyo International Airport Authority
The details of the financial resources for the New Tokyo International Airport Authority are provided in authority data. Local expenditures are not utilized as a financial resource by New Tokyo International Airport Authority.

3) Kansai International Airport Corporation
The details of the financial resources for the Kansai International Airport Corporation are provided in the "Airport Handbook."

4) Other Airports
The local airport maintenance expense of other airport is appropriated as a burden on the local governments.

Calculations of total investment and financial resources composition for airport improvement are shown in Table 5.3(1).

Tables 5.1–5.3 provide the calculation results for roadways, railways and airports. The numerical values are given as a percentage of GDP in the Tables and are converted into U.S. dollars in the Appendix. The results are also illustrated and discussed in Chapter 7.

Table 5.3(1). Investment amounts for airport improvements and financial resource composition.

		Investment amount	National expenditures	Local expenditures	Owners / Operators' funds	Debt	New construction	Maintenance and repaire
Germany	Munich	1,063	26	74	400	563	100.00%	0.00%
France	Total	2,790	376	169	1,217	1,027		
	Aéroport de Paris (ADP)	1,725	22	0	953	750		
	Others	1,065	354	169	264	277		
United Kingdom	BAAplc	219	0	0	219	0		
Unites States	Total	17,192	8,654	0	7,775	763	49.34%	50.66%
	Federal Aviation Administration	6,754	6,754	0	0	0		
	Others	10,438	1,900	0	7,775	763		
Japan	Total	765	139	29	233	365	84.89%	15.11%
	New Tokyo International Airport	88	23	0	11	55		
	Kansai International Airport (KIX)	233	52	13	9	159		
	Others	444	65	16	212	151		

Unit, Year	Germany	Million DM	-
	France	Million FF	1993
	U.K.	Million £	1991
	U.S.A.	Million US$	1992
	Japan	Billion Yen	1992

Table 5.3(2). Investment amounts for airport improvements and financial resource composition (GDP ratio).

	Investment amount	National expenditures	Local expenditures	Owners / Operators' funds	Debt
Germany	0.038%	0.001%	0.003%	0.014%	0.020%
France	0.039%	0.005%	0.002%	0.017%	0.014%
United Kingdom	0.038%	0.000%	0.000%	0.038%	0.000%
United States	0.286%	0.144%	0.000%	0.129%	0.013%
Japan	0.165%	0.030%	0.006%	0.050%	0.079%

Germany : Values of Munich International Airport
U.K.　　 : Values of BAAplc

Chapter 6

The calculation of contributors' shares

Earlier investment was classified as National Expenditure, Local Expenditures, Owners/operators' Funds and Debt but to complete the calculations of contributors' shares we must take account of taxes, rental fees and reimbursement of debt. The calculations in this chapter are based on the method discussed in Chapter 4.

6.1 Methods of calculation

Contributors' shares are divided into the following eight categories:

A) Payers of National Taxes
People burdened with the general funds of central or federal governments, such as income tax and corporate tax payers.

B) Payers of Local Taxes
People burdened with the general funds of local governments, such as residence tax payers.

C) Users
People burdened for utilizing the transportation systems who can be subdivided into:

C-1) Users (General taxes)
Users burdened with taxes from which revenues are allocated for general funds.

C-2) Users (Specific taxes, same mode)
Users burdened with taxes from which revenues are allocated to specific funds for same transportation mode.

C-3) Users (Specific taxes, other mode)
Users burdened with taxes from which revenues are allocated to specific funds for other transportation mode.

C-4) Users (Fees)
Users burdened with fees charged for usage of transportation systems.

D) Indirect Beneficiaries
People burdened with charges based on indirect benefits such as increase of property value.

E) Users (Later years)
Future users burdened with the reimbursement of debt.

Below, the previously calculated national and local expenditures, owners/operators' funds and debt are recalculated in consideration of the above-mentioned contributors.

6.2 ROADWAYS

For roadways, the national and local expenditures of each country are recalculated as shown in Table 6.1(1),(2). In this calculation, special attention was given to the various systems of specific funds for road improvement.

6.2.1 Germany

Germany has a system of specific funds for road improvement, so the actual payer of general funds and specific funds in the national and local expenditures were calculated.

National Expenditures are covered by revenues from automobile users' taxes. The amount is DM 9,859 million as shown in Table 6.1(1) and falls into the category of Users (Specific taxes, same mode).

Local expenditure (Table 6.1(2)) of DM 25,090 million is resourced by revenues from petroleum tax and automobile tax. It is made up of DM 12,047 million from specific funds[1], corresponding to Users (Specific taxes, same mode) and DM 13,043 million from general funds.

DM 104 million from local governments' general funds is created by automobile taxes[2]. The burden to Payers of Local Taxes is DM 12,939 million.

6.2.2 France

Automobile users' taxes are not placed in specific funds of road improvement so all national and local expenditure comes from general funds which were calculated as Payers of National Taxes, Payers of Local Taxes and Users (General taxes).

Table 6.1(1). Refigured amounts of national expenditures for roadways.

	General funds			Specific funds				Grand Total
	Payers of National Taxes	Users	Total	Payers of National Taxes	Users Same mode	Other modes	Total	
Germany	0	0	0	0	9,859	0	9,859	9,859
France	10,007	1,033	11,040	0	0	0	0	11,040
U.K.	1,618	339	1,958	0	0	0	0	1,958
U.S.A.	1,981	12	1,993	0	13,530	0	13,530	15,523
Japan	523	1	524	0	2,686	0	2,686	3,210

	General funds			Specific funds				Grand Total
	Payers of National Taxes	Users	Total	Payers of National Taxes	Users Same mode	Other modes	Total	
Germany	0.0%	0.0%	0.0%	0.0%	100.0%	0.0%	100.0%	100.0%
France	90.6%	9.4%	100.0%	0.0%	0.0%	0.0%	0.0%	100.0%
U.K.	82.7%	17.3%	100.0%	0.0%	0.0%	0.0%	0.0%	100.0%
U.S.A.	12.8%	0.1%	12.8%	0.0%	87.2%	0.0%	87.2%	100.0%
Japan	16.3%	0.0%	16.3%	0.0%	83.7%	0.0%	83.7%	100.0%

[1]   The proportion of automobile tax invested in road improvement differs in each state. The total automobile tax revenue is DM 13,317 million (value of Kraftfahrzeugsteuer in "Verkehr in Zahlen") of which 80% i.e. DM 10,654 million, goes into specific funds. DM 1,393 million was allocated as a subsidy for local road improvement from petroleum tax revenues (when computing at 2.7 penny/liter). Therefore, local expenditure from specific funds is DM 12,047 million.

[2]   The rate (α) of automobile tax revenue to the general accounts revenue in the local governments is calculated as follows:
α = total automobile tax revenue (value of Kraftfahrzeugsteuer in "Verkehr in Zahlen") $\times 0.2$/{general accounts revenue in local government (value of Länder, Gemeinden of Steuern und steuerähnliche Abgaben in "Statistisches Jahrbuch")—portion corresponding to the specific funds of automobile tax ($0.8 \times$ value of Kraftfahrzeugsteuer in "Verkehr in Zahlen")}.

Table 6.1(2). Refigured amounts of local expenditures for roadways.

	General funds			Specific funds				Grand Total
	Payers of Local Taxes	Users	Total	Payers of Local Taxes	Users Same mode	Other modes	Total	
Germany	12,939	104	13,043	0	12,047	0	12,047	25,090
France	36,845	8,756	45,601	0	0	0	0	45,601
U.K.	2,345	492	2,837	0	0	0	0	2,837
U.S.A.	20,719	327	21,046	0	27,691	0	27,691	48,737
Japan	4,522	216	4,738	0	2,112	0	2,112	6,851

	General funds			Specific funds				Grand Total
	Payers of Local tax	Users	Total	Payers of Local tax	Users Same mode	Other modes	Total	
Germany	51.6%	0.4%	52.0%	0.0%	48.0%	0.0%	48.0%	100.0%
France	80.8%	19.2%	100.0%	0.0%	0.0%	0.0%	0.0%	100.0%
U.K.	82.7%	17.3%	100.0%	0.0%	0.0%	0.0%	0.0%	100.0%
U.S.A.	42.5%	0.7%	43.2%	0.0%	56.8%	0.0%	56.8%	100.0%
Japan	66.0%	3.2%	69.2%	0.0%	30.8%	0.0%	30.8%	100.0%

Unit, Year			
	Germany	Million DM	1992
	France	Million FF	1993
	U.K.	Million £	1991
	U.S.A.	Million US$	1992
	Japan	Billion Yen	1992

Within National Expenditures, fuel tax and annual ownership tax produce a revenue of FF 1,033 million[3]. Therefore, the burden purely to Payers of National taxes becomes FF 10,007 million.

Within local expenditures, registration tax and annual ownership tax produce a revenue of FF 8,756 million[4]. Therefore, the burden to Payers of Local Taxes is FF 36,845 million.

---

[3] The users burden for general funds in the central government is calculated utilizing the rate (α) of the automobile users' tax revenue to the general accounts revenue.

α = {axle tax + fuel tax + special tax for corporate vehicles (figures taken from Annual Ownership and Other Taxes, Motor Fuel and Other Special Taxes of "World Road Statistics")}/general accounts revenue of the central government (figures taken from Les Recettes du Budget Général in "Annuaire Statistique de la France").

[4] The users burden for general funds of local governments are calculated from the rate (α) of the automobile users' tax revenues to the general accounts revenue.

α = {driver's license tax + registration tax + automobile tax (figures taken from Driver's License Fees, Tax on Acquisition and Annual Ownership Taxes for Cars of "World Road Statistics")}/general accounts revenue of the local government (figures taken from Impôts Directs Locaux of Impôts Directs par Nature d'impôt Versements Effectués ou rôles émis in "Annuaire Statistique de la France").

6.2.3 United Kingdom

As in France, there is no system of specific funds for road improvement in the United Kingdom so national and local expenditures come from general funds which include taxes imposed on automobiles users. Therefore the contents of general funds are calculated as Payers of National Taxes, Payers of Local Taxes and Users (General taxes).

National Expenditures from general funds, which include fuel tax and car registration tax, amount to £339 million[5]. Therefore, the burden for Payers of National Taxes is £1,618 million.

Automobile users do not contribute to local general funds. However, in the United Kingdom, the finances of the local governments depend on grants (RSG) from the central government in many cases and the portion corresponding to Users (General taxes) is calculated[6]. The amount is £492 million, and the burden purely for Payers of Local Taxes is £2,345 million.

6.2.4 United States

As in Germany, the United States has a system of specific funds for road improvement, so the actual payer of general funds and specific funds for national and local expenditures are calculated.

For National Expenditures from specific funds the revenues are from automobile fuel tax, road use tax and tyre tax. The amount of specific funds in Table 6.1(1) is U.S.$ 13,530 million[7]. All of the funds correspond to Users (Specific tax, same mode).

---

[5] The user burden for general funds of the central government are calculated by the rate (α) of automobile users' tax revenue to the general account revenue.

$\alpha = \{$automobile tax + fuel tax + motorcar tax (figures taken from Annual Ownership Taxes, Motor Fuel and Taxes on Acquisition of "World Road Statistics")$\}$/general accounts revenue of the central government (figures taken from Tax Revenue Current Receipts in "Annual Abstract of Statistics").

[6] The user burden for general funds of the local governments is calculated from the rate (α) of the automobile users' tax revenues to general accounts revenue of the central government.

[7] Automobile users' taxes are transferred to the Highway Trust Fund, and most are allocated to road improvement. The amount allocated to road improvement from the Highway Trust Fund is shown as the value Net Used for Highway Purposes of Funding for Highways and Disposition of Highway-User Revenues in "Highway Statistics." Coordination of the values for revenue and investment amounts totals U.S.$ 13,530 million.

110

The amount of general funds is U.S.$ 1,993 million. This includes U.S.$ 12 million automobile users taxes paid into general funds[8]. Therefore, the burden by Payers of National Taxes is U.S.$ 1,981 million.

For local expenditures, revenues from automobile tax and driver's license tax go into specific funds. The amount corresponding to specific funds in Table 6.1 (2) is U.S.$ 27,691 million[9]. All of the funds correspond to Users (Specific taxes, same mode).

The amount of general funds is U.S.$ 21,046 million. This includes U.S.$ 327 million of automobile users taxes[10]. Therefore, the burden by Payers of Local Taxes is U.S.$ 20,719 million.

6.2.5 Japan

Like Germany and the United States, Japan also has a system of specific funds for road improvement. Therefore, the actual payer of general funds and specific funds for national and local expenditures were calculated.

---

[8] The rate (α) of automobile users' tax revenue to general accounts revenue of the federal government is calculated as follows.
α = portion corresponding to general funds provided by automobile users' tax revenue (figure taken from Amount for Non-highway Purposes and Amount for Territories of the section Funding for Highways and Disposition of Highway-User Revenues in "Highway Statistics")/general account revenue of the federal government (figured by taking the value from Federal Government Tax Revenue in "Annual Abstract of Statistics"—Net used for Highway Purposes in the section Funding for Highways and Disposition of Highway-User Revenues in "Highway Statistics").

[9] The figure stated as Net Used for Highway Purposes of the section Funding for Highways and Disposition of Highway-User Revenues in "Highway Statistics" is the amount allocated for road improvements out of automobile users' tax revenues as well as the case of the federation. Combining the values for revenue and investment amounts, the total becomes U.S.$ 27,691 million.

[10] The rate (α) of automobile users' tax revenues to general accounts revenue in the state and local governments was calculated as follows.
α = portion corresponding to general funds provided by automobile users' tax revenue (figures taken from Amount for Non-Highway Purposes and Amount for Collection Expenses of the section Funding for Highways and Disposition of Highway-User Revenues in "Highway Statistics")/general account revenues of state and local governments (figures taken from State and Local Government Tax Revenues in "Annual Abstract of Statistics"— Net Used for Highway Purposes in the section Funding for Highways and Disposition of Highway-User Revenues in "Highway Statistics").

For National Expenditures, revenues from gasoline and petroleum gas taxes go into specific funds. The specific funds as shown in Table 6.1(1) becomes ¥2,686 billion[11]. All correspond to Users (Specific taxes, same mode). The amount of general funds is ¥524 billion. This includes automobile tonnage tax of ¥1.3 billion[12]. Therefore, the burden of Payers of National Taxes is ¥523 billion.

Specific funds for local expenditures are created from gas/oil delivery tax, automobile acquisition tax, etc. The total of specific funds in Table 6.1(2) becomes ¥2,112 billion[13]. All funds correspond to Users (Specific taxes, same mode).

Total general funds are ¥4,738 billion. This includes ¥216 billion from automobile tax and light vehicle tax[14]. Therefore, the burden by Payers of Local Taxes is ¥4,522 billion.

---

[11] Of the automobile users' taxes in Japan, the revenues from the gasoline tax and 1/2 of the revenues from the petroleum gas tax are allocated to the central government's specific funds for road improvement. Of the automobile tonnage tax, 80% of the 3/4 portion provided is normally utilized as specific funds. Specific funds for National Expenditures = gasoline tax + petroleum tax × (1/2) + automobile tonnage tax × (3/4) × (4/5) (from "Financial Statistics").

[12] The rate (α) of automobile users' tax revenue to general accounts revenue of the central government is calculated as follows:
α = automobile tonnage tax × (3/4) × (1/5)/{national tax—(gasoline tax + petroleum gas tax + automobile tonnage tax × (3/4) × (4/5) + local road tax)} (from "Financial Statistics").

[13] Local governments' specific funds come from: gas/oil delivery tax, automobile acquisition tax, automobile tonnage transferred tax, petroleum gas transferred tax and local road transferred tax. The automobile tonnage transferred tax is 1/4 of the automobile tonnage tax, the petroleum gas transferred tax is 1/2 of the petroleum gas tax and the local road transfer tax is a local road tax. Specific funds for local expenditures = gas/oil delivery tax + automobile acquisition tax + automobile tonnage transferred tax + petroleum gas transferred tax + local road transferred tax (from "Financial Statistics").

[14] The rate (α) of automobile users' tax revenue to general account revenue of the local governments is calculated as follows:
α = (automobile tax + light vehicle tax (from "Financial Statistics"))/{local tax ("Reference Calculation Data for Local Taxes")—(gas/oil delivery tax + automobile acquisition tax (from "Financial Statistics"))}.

6.3 RAILWAYS

National and local expenditures of each country are recalculated as shown in Table 6.2(1),(2) with special attention being given to the use of automobile users' tax revenues for railway improvements.

6.3.1 Germany

A portion of petroleum tax revenue is allocated as specific funds for the National Expenditures of public urban transportation. The amount of specific funds in Table 6.2(1) is DM 907 million[15]. All of the funds correspond to Users (Specific taxes, other mode).

The amount of general funds is DM 9,297 million, but none of this is burdened on railway users. Therefore, all of the funds become the burden of the Payers of National Taxes.

6.3.2 France

Taxes which are the burden of transportation users are not allocated as specific funds for railway improvement. There are no charges on railways users going to general funds, therefore, all National Expenditures are purely the burden of the Payers of National Taxes. The amounts are: FF 10,871 million for SNCF and FF 694 million for RATP, totaling FF 11,565 million.

For local expenditures, the "Versement de Transport," a charge on offices experiencing indirect benefits, is allocated as financial resources of the local governments for railway improvement and is sufficient to cover all local expendi-

---

[15] The petroleum tax allocated for public urban passenger transportation is DM 2,514 million (computed from "Outline of Railway Improvement in Foreign Countries"). The amount of petroleum tax allocated for the improvement of transportation systems as a portion of National Expenditures is calculated as follows: petroleum tax allocated for public urban passenger transportation (DM 2,514 million) × National Expenditures (DM 10,204 million)/{operating grants from the federal government (DM 9,600 million) + capital grants from the federal government (DM 18,688 million)}.

Table 6.2(1). Refigured amounts of national expenditures for railways.

	General funds			Specific funds				Grand Total
	Payers of National Taxes	Users	Total	Payers of National Taxes	Users Same mode	Other modes	Total	
Germany	9,297	0	9,297	0	0	907	907	10,204
France	11,565	0	11,565	0	0	0	0	11,565
U.K.	975	0	975	0	0	0	0	975
U.S.A.	951	0	951	0	0	820	820	1,771
Japan	123	0	123	0	0	0	0	123

	General funds			Specific funds				Grand Total
	Payers of National Taxes	Users	Total	Payers of National Taxes	Users Same mode	Other modes	Total	
Germany	91.1%	0.0%	91.1%	0.0%	0.0%	8.9%	8.9%	100.0%
France	100.0%	0.0%	100.0%	0.0%	0.0%	0.0%	0.0%	100.0%
U.K.	100.0%	0.0%	100.0%	0.0%	0.0%	0.0%	0.0%	100.0%
U.S.A.	53.7%	0.0%	53.7%	0.0%	0.0%	46.3%	46.3%	100.0%
Japan	100.0%	0.0%	100.0%	0.0%	0.0%	0.0%	0.0%	100.0%

Table 6.2(2). Refigured amounts of local expenditures for railways.

	General funds			Specific funds				Grand Total
	Payers of Local Taxes	Users	Total	Payers of Local Taxes	Users Same mode	Other modes	Total	
Germany	0	0	0	0	0	0	0	0
France	0	0	0	0	0	0	0	0
U.K.	0	0	0	0	0	0	0	0
U.S.A.	934	0	934	384	0	390	774	1,708
Japan	167	0	167	0	0	0	0	167

	General funds			Specific funds				Grand Total
	Payers of Local Taxes	Users	Total	Payers of Local Taxes	Users Same mode	Other modes	Total	
Germany	0.0%	0.0%	0.0%	0.0%	0.0%	0.0%	0.0%	0.0%
France	0.0%	0.0%	0.0%	0.0%	0.0%	0.0%	0.0%	0.0%
U.K.	0.0%	0.0%	0.0%	0.0%	0.0%	0.0%	0.0%	0.0%
U.S.A.	54.7%	0.0%	54.7%	22.5%	0.0%	22.8%	45.3%	100.0%
Japan	100.0%	0.0%	100.0%	0.0%	0.0%	0.0%	0.0%	100.0%

Unit, Year	Germany	Million DM	1993
	France	Million FF	1993
	U.K.	Million £	1993
	U.S.A.	Million US$	1993
	Japan	Billion Yen	1992

114

tures[16]. Therefore, all local expenditures can be appropriated as the burden by Indirect Beneficiaries.

6.3.3 United Kingdom

As in France, the tax burden on transportation users is not placed in specific funds for railway improvement. All national and local expenditures are general funds. There are charges on railways users going to general funds, therefore all National Expenditures are the burden of the Payers of National Taxes. The amounts are £595 million for BR and £380 million for LT, totaling £975 million.

6.3.4 United States

As in Germany, a portion of the automobile users' tax revenues is used to subsidize mass transport, so the actual payers of both general funds and specific funds for national and local expenditures are calculated. Such funds are not available to Class I Railroads or Amtrak. There are charges on railways users going into general funds, therefore all National Expenditures for Class I Railroads and Amtrak become purely the burden of the Payers of National Taxes. This amount is U.S.$ 485 million.

Regarding the National Expenditures for Transit, the portion of automobile users' taxes that go into specific funds (Table 6.2(1)) is U.S.$ 820 million[17]. This entire amount corresponds to Users (Specific taxes, other mode).

The amounts of general funds is U.S.$ 466 million. The burden is purely on the Payers of National Taxes.

---

[16] The amount collected by the "Versement de Transport" exceeds the amount of local expenditures calculated in Chapter 5 ("Outline of Railway Improvement in Foreign Countries"). There is no tangible data available showing the correct amount of grants to the SNCF or RATP by the "Versement de Transport".

[17] Funding for Highways and Disposition of Highway-User Revenues in "Highway Statistics" includes expenditures for mass transportation such as for busing. Therefore, the following computation was conducted to calculate the expenditures for railways (Transit): national government railway (Transit) expenditures = National Expenditures for Transit calculated in Chapter 5/federal total (operating funds, capital funds) in "Data Tables for the 1993 National Transit Database Section 15 Report" × amount for mass transportation (federal total from the Funding for Highways and Disposition of Highway-User Revenues in "Highway Statistics.")

The total National Expenditures burden on the Payers of National Taxes for Class I Railroads, Amtrak and Transit becomes U.S.$ 951 million, as shown in Table 6.2(1).

Regarding local expenditures for Transit, the portion of automobile users' taxes that goes into specific funds (Users (Specific taxes, other mode)) is U.S.$ 390 million[18]. Additionally, Income Taxes and Sales Taxes are dedicated as sources for transit operation and capital funding. These are specific funds, however, the actual contributors are the Payers of Local Taxes. The amount is U.S.$ 384 million[19].

6.3.5 Japan

Transport user taxes are not allocated as specific funds for railway improvement, so all national and local expenditures come from general funds. There are no charges on railways users going into general funds. Thus all national and local expenditures become purely the burden of the Payers of National Taxes and Payers of Local Taxes, respectively. These amounts are ¥123 billion and ¥167 billion. Additionally, in the same manner as for the former Japan National Railway, the burden of Payers of National Taxes becomes ¥34 billion.

6.4 AIRPORTS

For airports, the national and local expenditures for each country are recalculated as shown in Table 6.3(1),(2). Special attention should be given to the existence of systems with specific funds for airport improvement.

---

[18] The local railway (Transit) expenditures were calculated as mentioned in footnote 17: local expenditures for Transit calculated in Chapter 5/total state and local value (operating funds, capital funds) in "Data Tables for the 1993 National Transit Database Section 15 Report" × amount for mass transportation, total of states and local governments of Funding for Highways and Disposition of Highway-User Revenues in "Highway Statistics."

[19] Value = local expenditures for Transit calculated in Chapter 5 × (value of state and local taxes dedicated at source for applied transit operating funds and capital funds— value of gasoline taxes (taken from "Data Tables for the 1993 National Transit Database Section 15 Report")/total state and local value (operating funds, capital funds) in "Data Tables for the 1993 National Transit Database Section 15 Report.")

6.4.1 Germany

There is no system of specific funds for airport improvement in Germany. All national and local expenditures are general funds, and there are no charges on airport users going into general funds. Therefore, all national and local expenditures become purely the burden of Payers of National Taxes and Payers of Local Taxes. These amounts are DM 26 million and 74 million, respectively.

6.4.2 France

The central government of France has a system that utilizes specific funds for airport improvement called the Private Sector Aviation Special Account (Budget Annexe de l'Aviation Civile: BAAC). The financial resources include tolls and taxes from airways and users. The actual payers of the general funds and specific funds were calculated for National Expenditures[20]. As a result, Users (Specific taxes, same mode) is FF 146 million, and burden purely by Payers of National Taxes is FF 230 million.

All local expenditures come from general funds, and there are charges on airport users going into general funds. Therefore, all local expenditures are the burden of Payers of Local Taxes. The amount is FF 169 million.

6.4.3 United Kingdom

In the United Kingdom, the British Airport Authority Public Limited Company utilizes no airport improvement financial resources from national or local expenditures. Therefore, no additional calculations are required.

6.4.4 United States

The United States has a specific fund for airport improvement so both general funds and specific funds for National Expenditures must be calculated.

---

[20] Users (Specific taxes, same mode) = value of Subventions dont Etatin in "Activité des Aéroports Français Anné" and Payers of National Taxes = value of Subventions dont CEE in "Activité des Aéroports Français Anné."

Table 6.3(1). Refigured amounts of national expenditures for airports.

	General funds			Specific funds				Grand Total
	Payers of National Taxes	Users	Total	Payers of National Taxes	Users Same mode	Other modes	Total	
Germany	26	0	26	0	0	0	0	26
France	230	0	230	0	146	0	146	376
U.K.	0	0	0	0	0	0	0	0
U.S.A.	2,250	0	2,250	0	6,404	0	6,404	8,654
Japan	66	0	66	0	73	0	73	139

	General funds			Specific funds				Grand Total
	Payers of National Taxes	Users	Total	Payers of National Taxes	Users Same mode	Other modes	Total	
Germany	100.0%	0.0%	100.0%	0.0%	0.0%	0.0%	0.0%	100.0%
France	61.2%	0.0%	61.2%	0.0%	38.8%	0.0%	38.8%	100.0%
U.K.	0.0%	0.0%	0.0%	0.0%	0.0%	0.0%	0.0%	0.0%
U.S.A.	26.0%	0.0%	26.0%	0.0%	74.0%	0.0%	74.0%	100.0%
Japan	47.5%	0.0%	47.5%	0.0%	52.5%	0.0%	52.5%	100.0%

Table 6.3(2). Refigured amounts of local expenditures for airports.

	General funds			Specific funds				Grand Total
	Payers of Local Taxes	Users	Total	Payers of Local Taxes	Users Same mode	Other modes	Total	
Germany	74	0	74	0	0	0	0	74
France	169	0	169	0	0	0	0	169
U.K.	0	0	0	0	0	0	0	0
U.S.A.	0	0	0	0	0	0	0	0
Japan	15	0	15	0	13	0	13	29

	General funds			Specific funds				Grand Total
	Payers of Local Taxes	Users	Total	Payers of Local Taxes	Users Same mode	Other modes	Total	
Germany	100.0%	0.0%	100.0%	0.0%	0.0%	0.0%	0.0%	100.0%
France	100.0%	0.0%	100.0%	0.0%	0.0%	0.0%	0.0%	100.0%
U.K.	0.0%	0.0%	0.0%	0.0%	0.0%	0.0%	0.0%	0.0%
U.S.A.	0.0%	0.0%	0.0%	0.0%	0.0%	0.0%	0.0%	0.0%
Japan	53.8%	0.0%	53.8%	0.0%	46.2%	0.0%	46.2%	100.0%

Unit, Year	Germany	Million DM	-
	France	Million FF	1993
	U.K.	Million £	1991
	U.S.A.	Million US$	1992
	Japan	Billion Yen	1992

Germany : Values of Munich International Airport
U.K : Values of BAAplc

Specific funds for National Expenditures are created through ticket tax, aviation freight tax, etc, and total U.S.$ 6,404 million, as shown in Table 6.3(1)[21]. All the funds correspond to Users (Specific taxes, same mode). The amount of general funds is U.S.$ 2,250 million and is the burden of the Payers of National Taxes.

6.4.5 Japan

Like the United States, Japan also has a system of specific funds for airport improvement, so the actual payer of general funds and specific funds must be calculated for national and local expenditures.

National Expenditures from specific funds is ¥73 billion ("Financial Statistics"). All correspond to Users (Specific taxes, same mode). The amount of general funds is ¥66 billion and is the burden of Payers of National Taxes.

For local expenditures, the financial resource for specific funds is aircraft fuel transferred tax and amounts to ¥132 billion. All correspond to Users (Specific taxes, same mode). The amount of general funds is ¥15 billion and is the burden of Payers of Local Taxes.

In the above results, the actual contributors to national and local expenditures were calculated utilizing the categories of Payers of National Taxes, Payers for Local Taxes, User (General taxes), Users (Specific taxes, same mode) and Users (Specific taxes, other mode). Additionally, owners/operators' funds were classified as Users (Fees) and debt as Users (Later years). These classifications enabled the calculation of the burden of all contributors. The results are shown in Tables 6.4–6.6. In these tables, one each for roads, railways and airports, the shares of the actual contributors are shown in monetary terms and the GDP ratio of the country concerned. A table displaying the same values in U.S. dollars is provided in the Appendix.

---

[21] Operation (Airport and Airway Trust Fund) + grants-in-aid for airports + facilities and equipment (from FAA Statistical Handbook of Aviation).

Table 6.4(1). Shares of contributors for roadway improvements.

Country	FY	Payers of National Taxes	Payers of Local Taxes	Users (General taxes)	Users(Specific taxes)		Users (Fees)	Indirect Beneficiaries	Users (Later years)	Investment amounts
					Same mode	Other modes				
Germany	1992	0	12,939	104	21,906	0	0	0	0	34,949
France	1993	10,007	36,845	9,789	0	0	2,526	0	12,141	71,308
United Kingdom	1991	1,618	2,345	831	0	0	0	0	0	4,795
United States	1992	1,981	20,719	339	41,221	0	2,998	0	8,080	75,338
Japan	1992	523	4,522	217	4,798	0	937	0	1,818	12,815

*Germany: Million DM, France: Million FF, U.K.: Million £, U.S.A.: Million US$, Japan: Billion Yen

Table 6.4(2). Shares of contributors for roadway improvements (GDP ratio).

Country	FY	Payers of National Taxes	Payers of Local Taxes	Users (General taxes)	Users(Specific taxes)		Users (Fees)	Indirect Beneficiaries	Users (Later years)	Investment amounts
					Same mode	Other modes				
Germany	1992	0.000%	0.463%	0.004%	0.784%	0.000%	0.000%	0.000%	0.000%	1.251%
France	1993	0.141%	0.519%	0.138%	0.000%	0.000%	0.036%	0.000%	0.171%	1.005%
United Kingdom	1991	0.282%	0.409%	0.145%	0.000%	0.000%	0.000%	0.000%	0.000%	0.835%
United States	1992	0.033%	0.344%	0.006%	0.685%	0.000%	0.050%	0.000%	0.134%	1.251%
Japan	1992	0.113%	0.975%	0.047%	1.034%	0.000%	0.202%	0.000%	0.392%	2.763%

Table 6.5(1). Shares of contributors for railway improvements.

Country	FY	Payers of National Taxes	Payers of Local Taxes	Users (General taxes)	Users(Specific taxes)		Users (Fees)	Indirect Beneficiaries	Users (Later years)	Investment amounts
					Same mode	Other modes				
France	1993	11,565	0	0	0	0	4,489	1,092	3,374	20,521
United Kingdom	1993	975	0	0	0	0	1,207	0	162	2,344
United States	1993	951	1,318	0	0	1,210	6,172	0	0	9,651
Japan	1992	123	167	0	0	0	761	0	675	1,726
Germany (DB/DR)	1993	9,297	0	0	0	907	2,468	0	1,231	13,904
France (SNCF)	1993	10,871	0	0	0	0	3,476	551	1,914	16,812
United Kingdom (BR)	1993	595	0	0	0	0	960	0	162	1,717
United States (Class I, Amtrak)	1993	485	0	0	0	0	5,764	0	0	6,250
Japan (JNR)	1986	34	0	0	0	0	405	0	274	713

*Germany: Million DM, France: Million FF, U.K.: Million £, U.S.A.: Million US$, Japan: Billion Yen

Table 6.5(2). Shares of contributors for railway improvements (GDP ratio).

Country	FY	Payers of National Taxes	Payers of Local Taxes	Users (General taxes)	Users(Specific taxes)		Users (Fees)	Indirect Beneficiaries	Users (Later years)	Investment amounts
					Same mode	Other modes				
France	1993	0.163%	0.000%	0.000%	0.000%	0.000%	0.063%	0.015%	0.048%	0.289%
United Kingdom	1993	0.156%	0.000%	0.000%	0.000%	0.000%	0.193%	0.000%	0.026%	0.374%
United States	1993	0.015%	0.021%	0.000%	0.000%	0.019%	0.097%	0.000%	0.000%	0.152%
Japan	1992	0.026%	0.036%	0.000%	0.000%	0.000%	0.164%	0.000%	0.146%	0.372%
Germany (DB/DR)	1993	0.328%	0.000%	0.000%	0.000%	0.032%	0.087%	0.000%	0.043%	0.491%
France (SNCF)	1993	0.153%	0.000%	0.000%	0.000%	0.000%	0.049%	0.008%	0.027%	0.237%
United Kingdom (BR)	1993	0.095%	0.000%	0.000%	0.000%	0.000%	0.153%	0.000%	0.026%	0.274%
United States (Class I, Amtrak)	1993	0.008%	0.000%	0.000%	0.000%	0.000%	0.091%	0.000%	0.000%	0.099%
Japan (JNR)	1986	0.010%	0.000%	0.000%	0.000%	0.000%	0.120%	0.000%	0.081%	0.211%

121

Table 6.6(1). Shares of contributors for airport improvements.

Country	FY	Payers of National Taxes	Payers of Local Taxes	Users (General taxes)	Users(Specific taxes) Same mode	Other modes	Users (Fees)	Indirect Beneficiaries	Users (Later years)	Investment amounts
France	1993	230	169	0	146	0	1,217	0	1,027	2,790
United States	1992	2,250	0	0	6,404	0	7,775	0	763	17,192
Japan	1992	66	15	0	86	0	224	9	365	765
Germany (Munich)	-	26	74	0	0	0	400	0	563	1,063
France (ADP)	1993	0	0	0	22	0	953	0	750	1,725
United Kingdom (BAAplc)	1991	0	0	0	0	0	219	0	0	219
Japan (NRT)	1992	11	0	0	12	0	11	0	55	88
Japan (KIX)	1992	25	7	0	33	0	0	9	159	233

*Germany: Million DM, France: Million FF, U.K.: Million £, U.S.A.: Million US$, Japan: Billion Yen
* Japan (NRT): Values of New Tokyo International Airport, Japan(KIX): Values of Kansai International Airport

Table 6.6(2). Shares of contributors for airport improvements (GDP ratio).

Country	FY	Payers of National Taxes	Payers of Local Taxes	Users (General taxes)	Users(Specific taxes) Same mode	Other modes	Users (Fees)	Indirect Beneficiaries	Users (Later years)	Investment amounts
France	1993	0.003%	0.002%	0.000%	0.002%	0.000%	0.017%	0.000%	0.014%	0.039%
United States	1992	0.037%	0.000%	0.000%	0.106%	0.000%	0.129%	0.000%	0.013%	0.286%
Japan	1992	0.014%	0.003%	0.000%	0.019%	0.000%	0.048%	0.002%	0.079%	0.165%
Germany (Munich)	-	0.001%	0.003%	0.000%	0.000%	0.000%	0.014%	0.000%	0.020%	0.038%
France (ADP)	1993	0.000%	0.000%	0.000%	0.000%	0.000%	0.013%	0.000%	0.011%	0.024%
United Kingdom (BAAplc)	1991	0.000%	0.000%	0.000%	0.000%	0.000%	0.038%	0.000%	0.000%	0.038%
Japan (NRT)	1992	0.002%	0.000%	0.000%	0.003%	0.000%	0.002%	0.000%	0.012%	0.019%
Japan (KIX)	1992	0.005%	0.001%	0.000%	0.007%	0.000%	0.000%	0.002%	0.034%	0.050%

* Japan (NRT): Values of New Tokyo International Airport, Japan(KIX): Values of Kansai International Airport

CHAPTER 7

COMPARATIVE CONSIDERATIONS

7.1 COMPARISON OF INVESTMENT AMOUNTS

In this chapter we make a cross-country comparison of total investment and actual contributors' shares as calculated in Chapters 5 and 6. The latter is especially important as it analyses financial resources with respect to the rate of general funds utilization, users' burden and the rate of debt.

7.1.1 Total investment

Total investment in roads, rail, and airports in each country has been reported in Tables 5.1–5.3. In Figure 7.1(1)–(3) those values are expressed as a

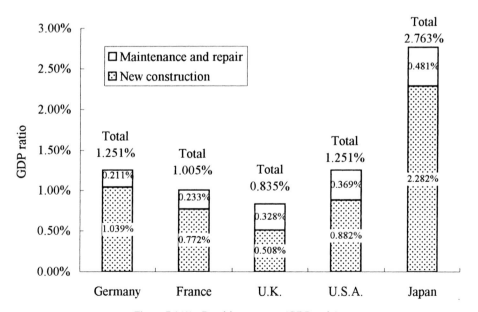

Figure 7.1(1). Road investments (GDP ratio).

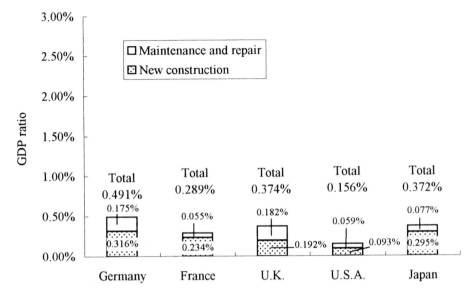

Figure 7.1(2). Railway investments (GDP ratio).

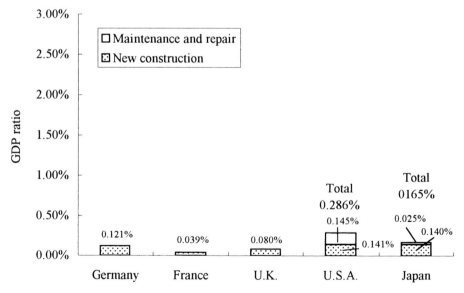

Figure 7.1(3). Airport investments (GDP ratio).

ratio of GDP, each broken down into its constituent components of new construction and maintenance. Figure 7.2(1)–(3) show trends in investment for five countries. Note that the total values vary slightly in regard to systems included (see figure legends).

Road investment as a proportion of GDP in the European countries and the U.S.A. is in the range 0.84–1.25% (Figure 7.1(1)), whereas Japan has a higher level at 2.763%. Japan's greater investment is partly due to the existence of specific fund and toll road systems.

The rates of maintenance and repair expenses are high in the United Kingdom and the United States, highlighting a general trend to shift investment away from new construction. As shown in Figure 7.1(2), railway investment in Germany (DB/DR) is quite a bit larger (0.491%) than those of other countries. After Germany, the highest levels are those of Japan (0.382%) and the United Kingdom (0.374%), while that of France is rather low (0.289%). The amount of investment in the United States is also low (0.156%).

Regarding airport improvements (Figure 7.1(3)), the United States shows a very large amount of improvement work (0.286%) as compared to Germany (0.121%), France (0.039%), the United Kingdom (0.038%), and Japan (0.165%), though data is limited for the United Kingdom and Germany. Note that for

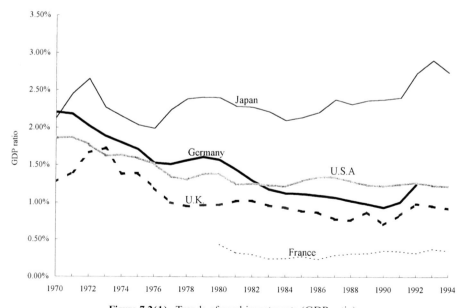

Figure 7.2(1). Trends of road investments (GDP ratio).

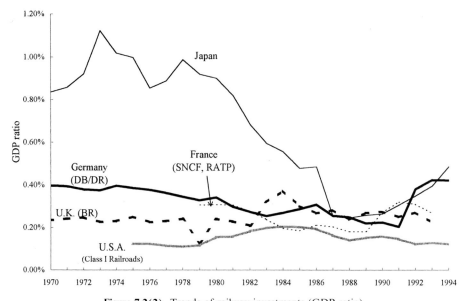

Figure 7.2(2). Trends of railway investments (GDP ratio).
*Germany: Values of DB/DR. Not including urban railways such as the U-Bahn.
*France: Values of SNCF and RATP. Not including urban railways except those in Paris.
*U.K.: Values of BR. Not including urban railways.
*U.S.A.: Values of Class I Railroads. Not including AMTRAK and urban railways.

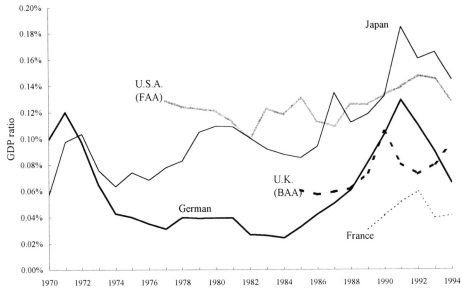

Figure 7.2(3). Trends of airport investments (GDP ratio).

each country, road improvement is the largest investment, followed by railway improvement. The exception is the United States, where more is invested in airports than in railways.

7.1.2 Composition of financial resource

The graphs in Figure 7.3(1)–(3) show the composition of financial resources by transport mode for investment as a ratio of GDP: Figure 7.4 (1) and (2) relate to national railways (or former national railways) and major airports.

For road improvement, local expenditures in Japan are very large (GDP ratio of 1.477%) as shown in Figure 7.3(1), and exceed the rates of total investment by European countries and the United States (0.84–1.25%). The utilization of owners/operators' funds and debt are, in descending order, Japan, France and United States. This result reflects a substantial dependence on toll roads in these countries. On the other hand, for railway improvement, Germany has nearly the same proportion of National Expenditures (0.36%) as the total investment amount of Japan (0.382%) and the United Kingdom (0.374%), and exceeds the total investment amounts of the other countries (0.102–0.289%). For airport improvements National Expenditures and owners/operators' funds in the

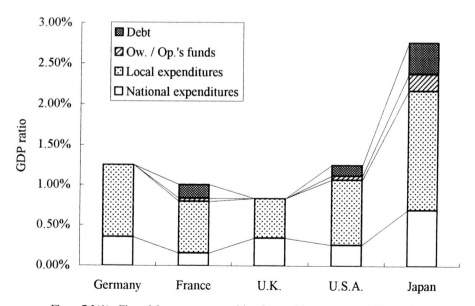

Figure 7.3(1). Financial resource composition for road improvements (GDP ratio).

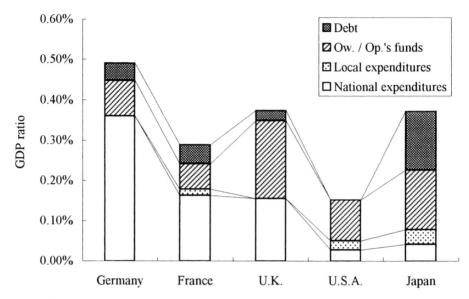

Figure 7.3(2). Financial resource composition for railway improvements (GDP ratio).

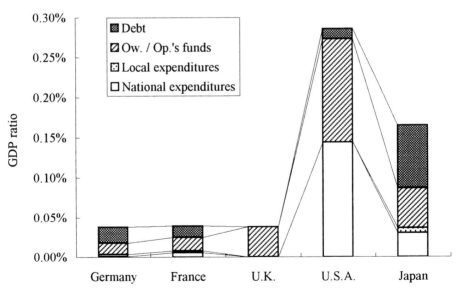

Figure 7.3(3). Financial resource composition for airport improvements (GDP ratio).

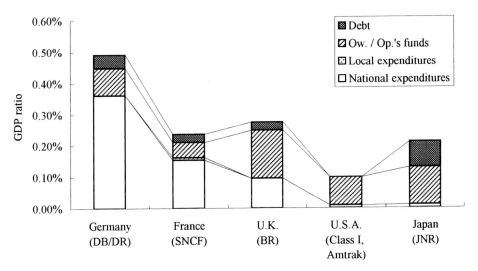

Figure 7.4(1). Financial resource composition for national railways (GDP ratio).

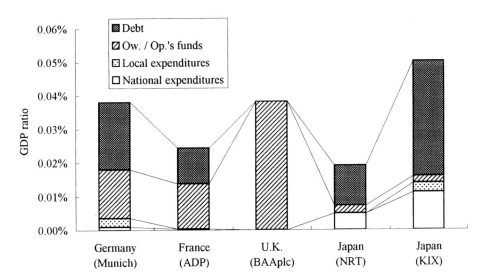

Figure 7.4(2). Financial resource composition for main airports (GDP ratio).

United States are the highest, followed by Japan. In the other countries, the National Expenditures are extremely small. In the case of national railway resources (Figure 7.4(1)) there is a wide variance across countries, with the National Expenditures in Germany and France being relatively high and those in the United States and Japan being relatively low. Note the high percentage of debt in Japan, both for the national railways and for the Kansai International Airport (Figures 7.4(1) and (2), respectively).

The graphs in Figure 7.5(1)–(3) use the values of Figure 7.3(1)–(3) rewritten as ratios of each component to total investment amount. It can be seen that the local expenditures for road improvement are large in every country. Another characteristic point is that owners/operators' funds and debt are high in France and Japan (accounting for approximately 20% of total investment amount), where toll road systems are currently being introduced. But looking at railway improvement, in Germany and France the totals for the central government and local expenditures rise above 50%, and in the United Kingdom and the United States they are 41.6% and 32.4%, respectively. Yet, in Japan the value is a low 16.32%. As discussed in Chapter 3, this disparity reflects a difference in national outlooks regarding the role of the government in improving railways. Also, in the case of airport improvement, in contrast to that of the railways, financial resources are nearly covered in Germany and France by owners/operators' funds and debt, whereas in Japan public funds are utilized. In the

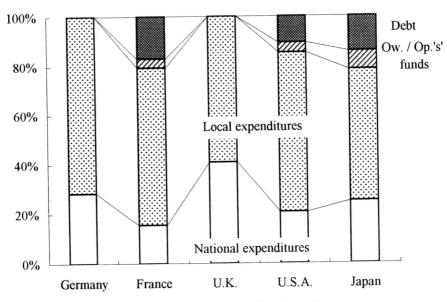

Figure 7.5(1). Financial resource composition for road improvements.

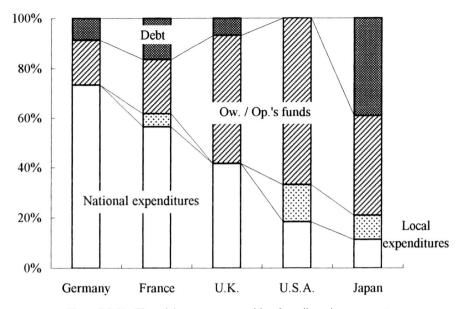

Figure 7.5(2). Financial resource composition for railway improvements.

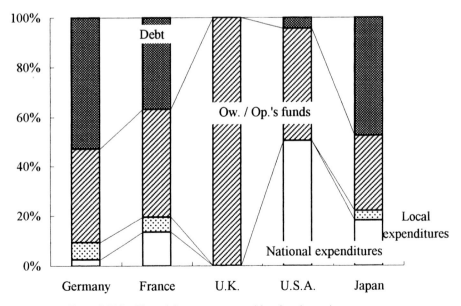

Figure 7.5(3). Financial resource composition for airport improvements.

132

United States, 50.3% is covered by public funds, the highest rate among all of the objects calculated.

7.2 SHARES OF CONTRIBUTORS

The calculation of contributors' shares (Tables 6.4–6.6) is an important characteristic of this book, and particularly for their use as an index of comparison. Graphs illustrating the values as a ratio to GDP are given in Figure 7.6(1)–(3). In addition, Figure 7.7(1) and (2) show the shares of contributors to national railways (or former national railways) and major airports. The results for road improvement (Figure 7.6(1)) show that the existence or non-existence of specific funds is reflected in the shares, i.e. the burden to users is larger in Germany, the United States, and Japan where there are specific funds, and lower in France and the United Kingdom where there are not, though there is a considerable amount of automobile users' taxes in general funds. A comparison of the shares of Payers of General Taxes shows that these shares are low in Germany and the United States. In Figure 7.3(1), National and Local Expenditures are not small, but a good portion of the expenses of the National and Local government are actually covered by automobile users' taxes, and the burden to general tax revenues is small, as shown in Figure 7.6(1). In terms of

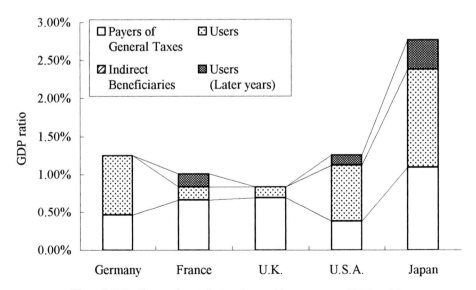

Figure 7.6(1). Shares of contributors for road improvements (GDP ratio).

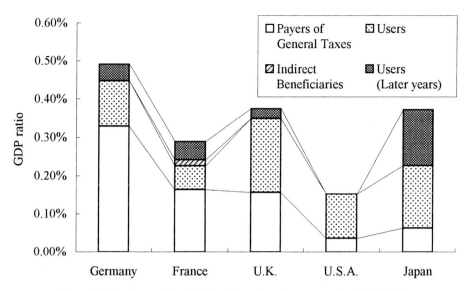

Figure 7.6(2). Shares of contributors for railways improvements (GDP ratio).
*Germany: Values of DB/DR. Not including urban railways such as the U-Bahn.
*France: Values of SNCF and RATP. Not including urban railways except those in Paris.
*U.K.: Values of BR and LT. Not including urban railways except those in London.

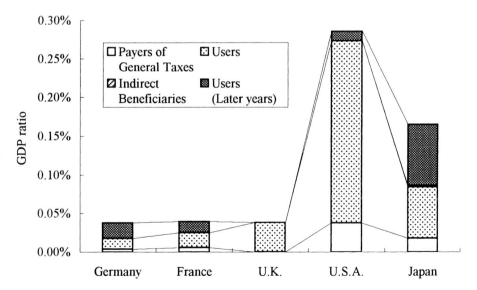

Figure 7.6(3). Shares of contributors for airport improvements (GDP ratio).

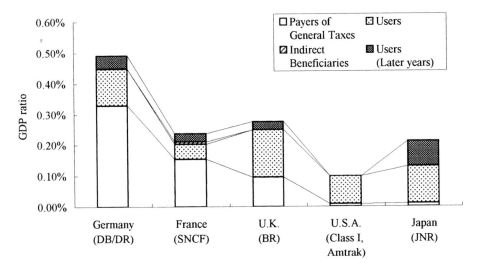

Figure 7.7(1). Shares of contributors for national railways (GDP ratio).

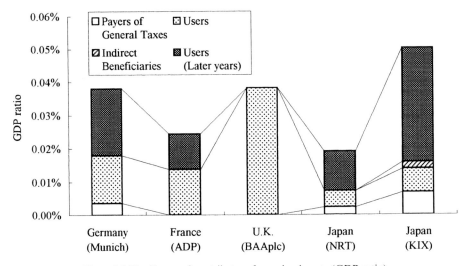

Figure 7.7(2). Shares of contributors for main airports (GDP ratio).

the burden placed on future generations, the values are largest for Japan, France, and the United States.

For railway improvement, the percentage of Payers of National Taxes in Germany is large. Specific taxes on users of other modes are seen in Germany and United States. The burden on users is large in every country. The burden to Users (Later years) is especially large in Japan.

For airport improvement, the percentages of both Users (Specific taxes, same mode) and Users (Fees) are large in the United States. Japan and the United States have a burden to Payers of National Taxes. Figure 7.3(3) demonstrates that the National Expenditures in the United States are more than 0.14%. However, the category of Payers of National Taxes in Figure 7.6(3) is approximately 0.04%. The graphs in Figure 7.8(1)–(3) compare contributors' ratios. For roadways, the shares of contributors are highly dependent on the existence or non-existence of a specific fund for road improvement. In Germany, the United States, and Japan, where specific funds are utilized, the burden placed on users, including Users (Specific taxes, same mode), accounts for 40–60% of the total. On the other hand, in France and the United Kingdom, which have no specific funds, the burden to Users (General taxes) accounts for some degree of the total (16.12–20.97%), but the burden to Payers of National Taxes and Payers of Local Taxes accounts for the largest portion of the total rates. In

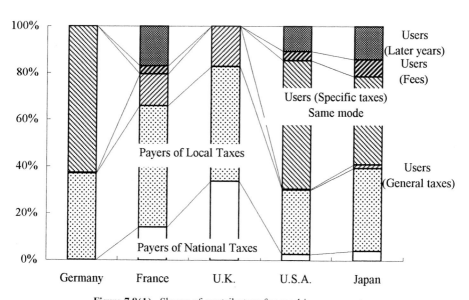

Figure 7.8(1). Shares of contributors for road improvements.

136

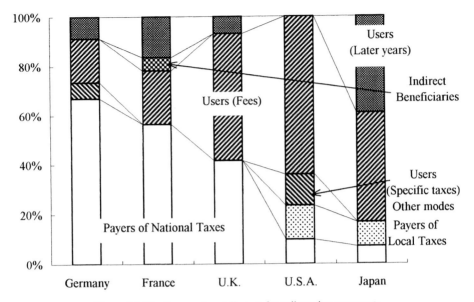

Figure 7.8(2). Shares of contributors for railway improvements.
*Germany: Values of DB/DR. Not including urban railways such as the U-Bahn.
*France: Values of SNCF and RATP. Not including urban railways except those in Paris.
*U.K.: Values of BR and LT. Not including urban railways except those in London.

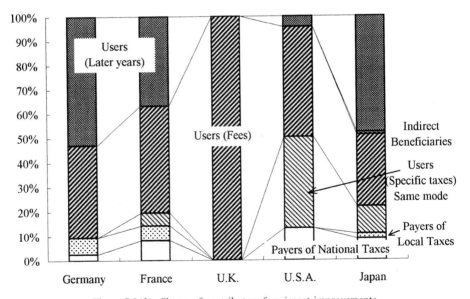

Figure 7.8(3). Shares of contributors for airport improvements.
*Germany: Value of Munich International Airport.
*U.K.: Value of BAAplc.

addition, the burden to Users (Later years) exceeds 10% in France, Japan, and the United States.

For railway improvement, the burden to Payers of National Taxes is more than 60% in Germany and exceeds 50% in France, but is less than 10% in the United States and Japan. The Users (Specific taxes, other mode), is a burden on road users, is also apparent in Germany and the United States, accounting for 6.5 and 8.7% of the total railway improvement burden, respectively. The Users (Later years) in Japan is a high 40.6%. For airport improvement, the burden to Payers of National Taxes accounts for a large rate in France, the United States, and Japan (8.24–13.09%) as compared to Germany and the United Kingdom. However, the burden to users created by Users (Specific taxes, same mode) and Users (Fees) is large as a whole. One characteristic is that the rate of Users (Later years) for airports is larger than that for roadways or railways.

7.3 CONSIDERATION BY COMPARATIVE GRAPHS

Figures 7.9–7.14 are provided in order to clearly show the differences in the contributors' shares of each country. Figure 7.9 is a graph illustrating the rate of burden on Payers of National Taxes and Payers of Local Taxes, and the rate of burden on the Users of every transportation mode in every country. Figure 7.10 is a graph showing the ratio of general funds in every country. The ratio of public funds is generally high for roadways and low for airports, but varies from country to country for railways. The graphs comparing the ratio of general funds and debt for railways and roadways are provided in Figure 7.11, 7.13, respectively. The graphs utilizing the GDP ratios are shown in Figure 7.12, 7.14. Figures 7.15–7.21 show the trends of shares of contributors. Regarding roads (Figure 7.15), the burden by public funds (i.e., Payers of National Tax and Local Tax) is increasing only in Japan. The burden by Users is decreasing in the United States and Germany, but not very significantly. Regarding railways (Figure 7.16), the burden by public fund is changing year by year. In general, this burden is largest in Germany and France. Regarding the trend for debt (Figure 7.18), Japan has the highest debt for both roads and railways.

7.4 CONTRIBUTORS' SHARES OF THE TOTAL BURDEN

The contributors' shares calculated in this book are determined by utilizing the amounts actually invested in a specific year. In addition to the total investment, there is also the reimbursement of debt due to investment in transportation

138

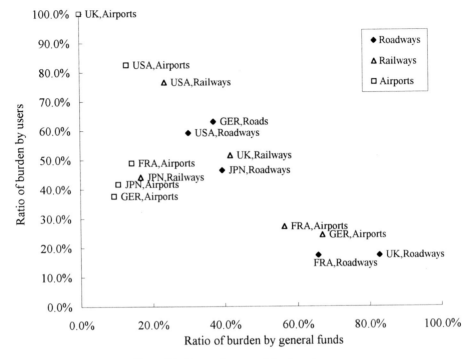

Figure 7.9. Burden by general funds and users.

improvement in previous years, which can be described by the following relation-ship: amount of burden in a year = amount invested in the year + amount of debt reimbursement in the year—amount of newly borrowed debt in the year. Therefore, if the amount of reimbursement in a year is larger than the new debt, the amount of actual burden in the year becomes larger than the investment amount; if smaller, the investment amount becomes larger than the burden amount. When considering such a relationship regarding contributors' shares, there are two methods of making the calculation. One is to calculate the shares for the amount invested in a year, and the other is to calculate the shares for the amount to be burdened within the year. Therefore, in addition to discussion of the contributors' shares for investment amounts, the shares of contributors to the burden amount are also calculated. In this case, out of the eight contributor categories, User (Later years) is removed and replaced with "Users burdened with reimbursement of debt." Figure 7.22(1)–(3) shows graphs that illustrate the contributors' shares for amounts actually burdened for transportation systems within the year as a ratio to GDP. Corresponding numerical values are provided in the Appendix.

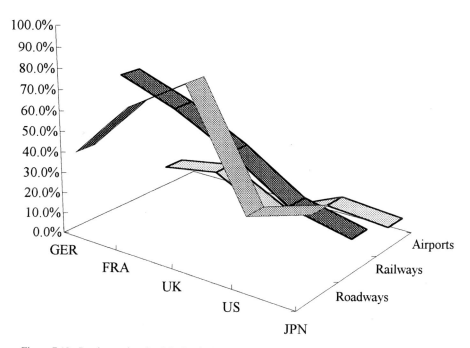

Figure 7.10. Burden ratio of public funds (payers of national taxes and payers of local taxes).

Regarding road improvement, for which the debt dependence is comparatively low, we observe nearly the same result as for the contributors' shares of investment amounts. However, there are some portions where the burden rates change for railways and airports. For example, when past investment burdens are imposed, this can cause the amount of burden to exceed that of investment. Likewise, in a situation when burden is transferred to the future, the burden amount is smaller for that particular year. For railways in France and Japan, the burden amounts are larger, and for the airports in every country, the investment amounts are larger. This reflects the past emphasis on railway investment and the present one on airport investment. However, the present generation's incurring of debt without burden may ultimately limit future investments.

140

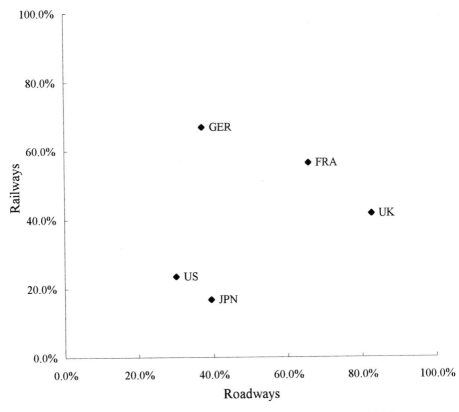

Figure 7.11. Comparison of burden by public funds (ratio): roadways and railways.

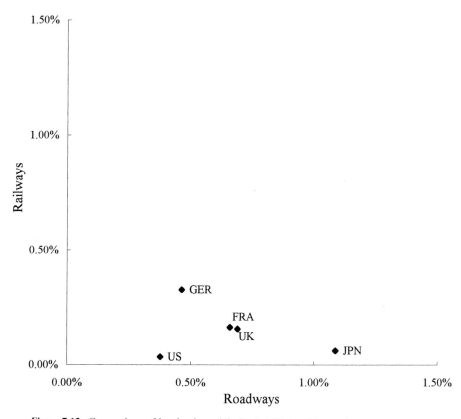

Figure 7.12. Comparison of burden by public funds (GDP ratio): roadways and railways.

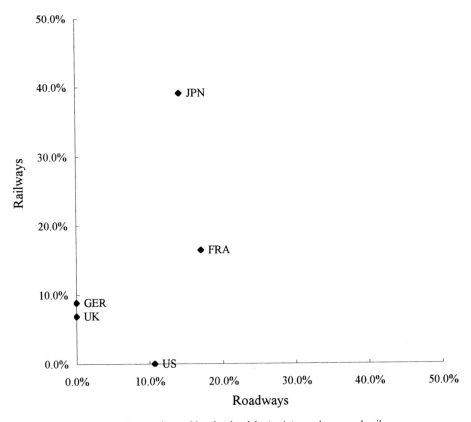

Figure 7.13. Comparison of burden by debt (ratio): roadways and railways.

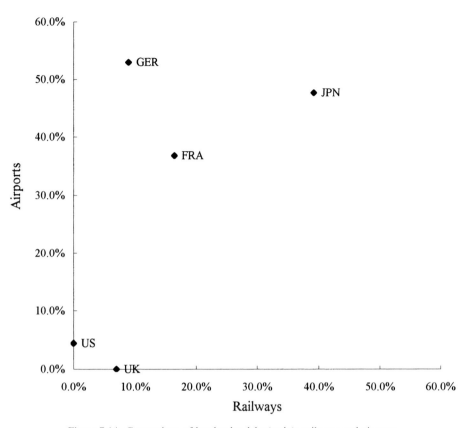

Figure 7.14. Comparison of burden by debt (ratio): railways and airports.

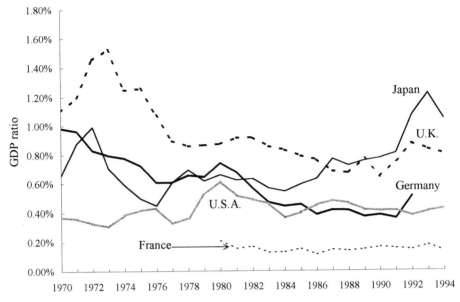

Figure 7.15(1). Trends of burden by public funds (GDP ratio)—roadways.

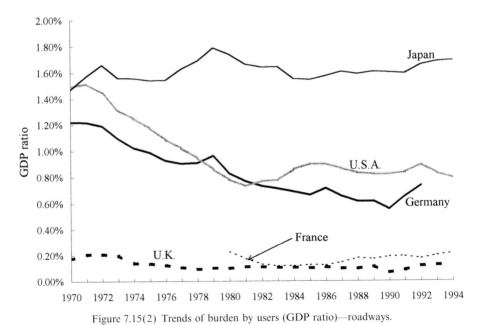

Figure 7.15(2) Trends of burden by users (GDP ratio)—roadways.
*France: Values of national highways and expressways. Not including local roads.

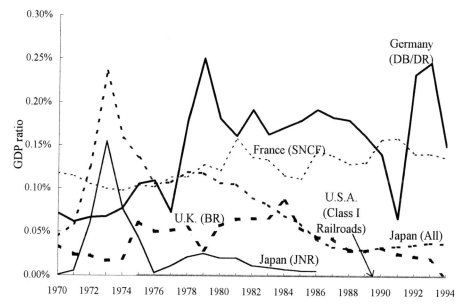

Figure 7.16(1). Trends of burden by public funds (GDP ratio)—railways.

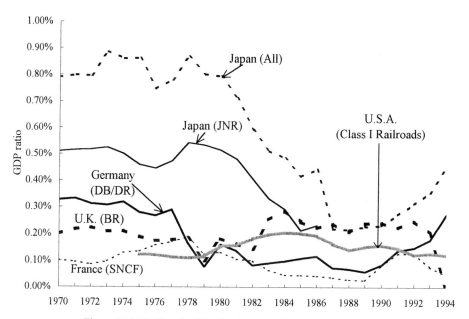

Figure 7.16(2). Trends of burden by users (GDP ratio)—railways.

*Germany: Values of DB/DR. Not including urban railways such as the U-Bahn.
*France: Values of SNCF. Not including urban railways.
*U.K.: Values of BR. Not including urban railways.
*U.S.A.: Values of Class I Railroads. Not including AMTRAK and urban railways.
*Japan (All): Values of all authorities.
*Japan (JNR): Values of Japan National Railways.

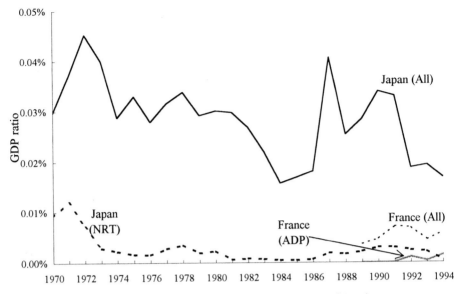

Figure 7.17(1). Trends of burden by public funds (GDP ratio)—airports.

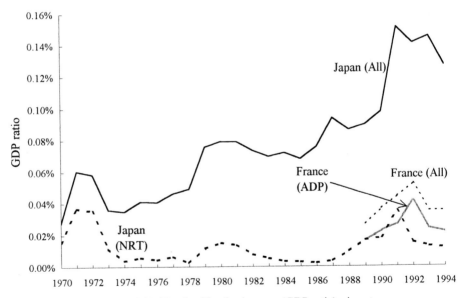

Figure 7.17(2). Trends of burden by users (GDP ratio)–airports.

*France (All): Values of all authorities.
*France (ADP): Values of ADP.
*Japan (All): Values of all authorities.
*Japan (NRT): Values of New Tokyo International Airport.

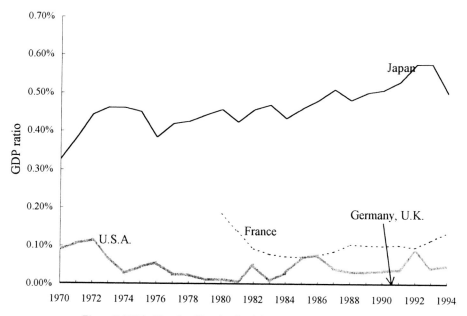

Figure 7.18(1). Trends of burden by debt (GDP ratio)—roadways.

*France: Values of national highways and expressways. Not including local roads.

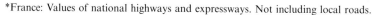
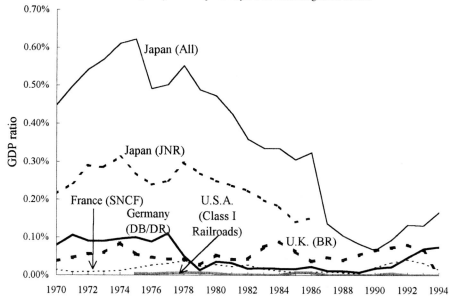

Figure 7.18(2). Trends of burden by debt (GDP ratio)—railways.

*Germany: Values of DB/DR. Not including urban railways such as the U-Bahn.
*France: Values of SNCF and RATP. Not including urban railways except those in Paris.
*U.K.: Values of BR. Not including urban railways.
*U.S.A.: Values of Class I Railroads. Not including AMTRAK and urban railways.
*Japan (All): Values of all authorities.
*Japan (JNR): Values of Japan National Railways.

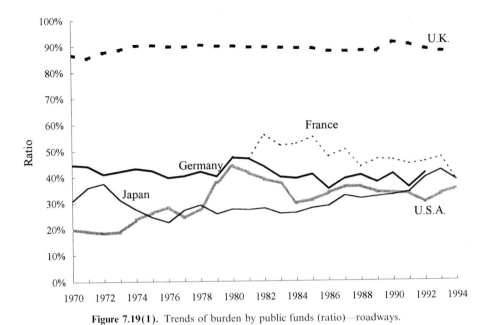

Figure 7.19(1). Trends of burden by public funds (ratio)—roadways.

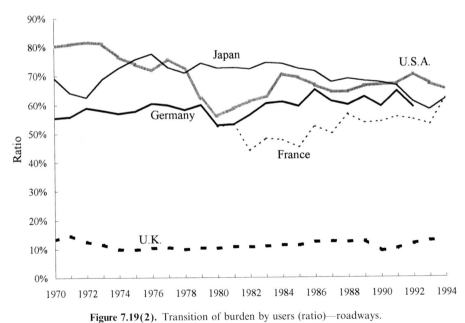

Figure 7.19(2). Transition of burden by users (ratio)—roadways.
France: Values of national highways and expressways. Not including local roads.

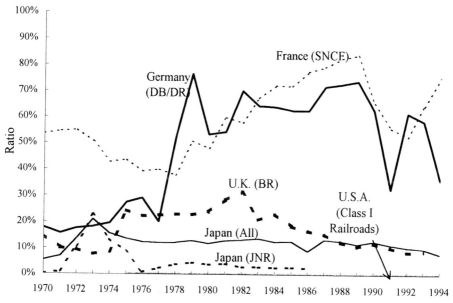

Figure 7.20(1). Trends of burden by public funds (ratio)—railways.

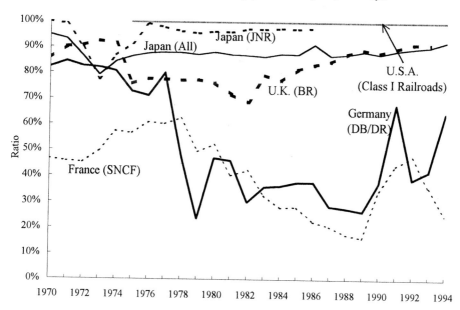

Figure 7.20(2). Trends of burden by users (ratio)—railways.

*Germany: Values of DB/DR. Not including urban railways such as U-Bahn.
*France: Values of SNCF. Not including urban railways.
*U.K.: Values of BR. Not including urban railways.
*U.S.A.: Values of Class I Railroads. Not including AMTRAK and urban railways.
*Japan (All): Values of all authorities.
*Japan (JNR): Values of Japan National Railways.

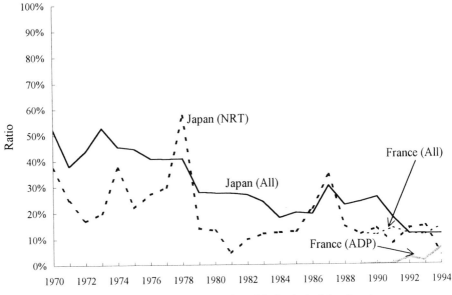

Figure 7.21(1). Trends of burden by public funds (ratio)—airports.

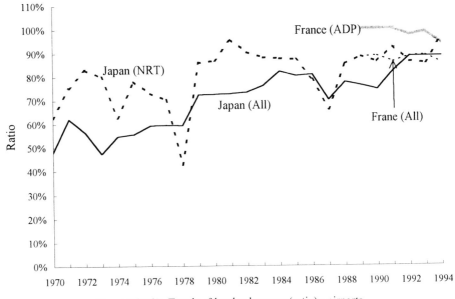

Figure 7.21(2). Trends of burden by users (ratio)—airports.

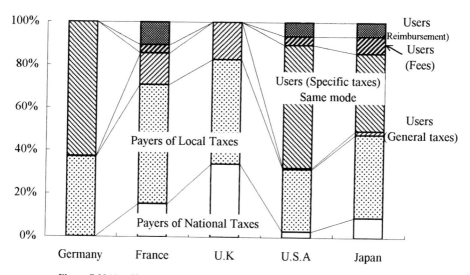

Figure 7.22(1). Shares of contributors for road improvements (burden ratio).

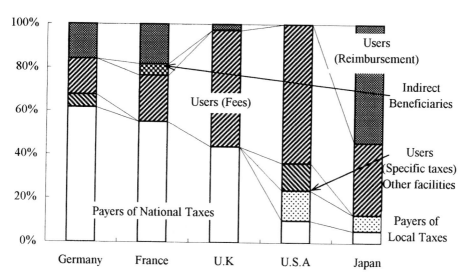

Figure 7.22(2). Shares of contributors for railway improvements (burden ratio).

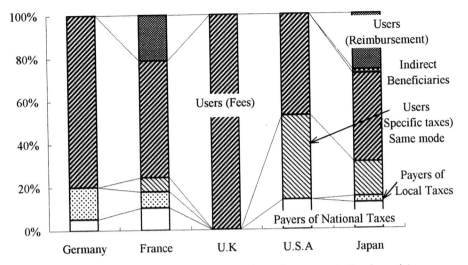

Figure 7.22(3). Shares of contributors for airport improvements (burden ratio).

*Germany: Value of Munich International Airport.
*U.K.: Value of BAAplc.

CHAPTER 8
PROSPECTS

Mankind has been building roads and exploring the seas since ancient times, gathering wisdom. The history of transportation improvement is the history of the advancement of mankind. The transportation infrastructure created over the years, which has required vast amounts of time and expenditures, continues to benefit our daily lives. It is incumbent upon us to continue this tradition by creating an optimal transportation infrastructure for the future. The following goals can be taken as requirements of future transportation systems worldwide.

(1) A balance between efficiency and social roles

As the available capital and financial solvency of individual countries continues to decline, governmental funding is an increasingly problematic means of improving transportation systems. Today, financial resources must be delegated judiciously and used efficiently.

The principle of privatization, which transfers ownership from the public to the private sector in order to utilize the latter's efficiency, will be increasingly relevant in the future.

However, it must also be understood that privatization does not always result in the most appropriate volume of investment or in appropriately shared burdens. As mentioned in Chapter 2, the existence of externalities and the issue of social justice must not be overlooked. If there is no charging system that covers the cost of protecting the environment and of promoting social welfare, privatization will fail. In many transportation projects, the total cost cannot be borne by the users, and in such cases, the project cannot be realized without the help of the public sector. In addition, there is a high possibility that the price of user charge decided by the private sector under the market mechanism will not lead to an optimal modal split.

Therefore, to maximize future improvements, individual projects will need to be evaluated not only by the profitability criteria of the private sector, but by the public criteria of social costs and benefits.

(2) Dependence on debt

In many countries, total debt in transport investment has grown to a formidable sum. It is, of course, fair that the burden for constructing an infrastructure that will benefit future generations should be distributed across both present and future generations, but the burden should not exceed the future benefit. If the future benefit is overestimated, the debt will be too large; sometimes future benefits are intentionally overestimated to defray the present burden. On the other hand, if the future benefit is underestimated, an insufficient investment will be made in the infrastructure, and in this case, the present generation will fail to fulfill its responsibility to improve the transportation infrastructure.

The important point to remember is that the present generation must decide this matter in absentia of the future beneficiaries, and therefore has a responsibility to evaluate the options carefully.

(3) Improvement as an international infrastructure

Another aspect of the problem of share imbalance among contributors is the increased utilization of transportation systems by nationals from countries not sharing the burden of cost. An effort must be made to coordinate the international imbalance of burden caused by this geographical spillover of benefits.

(4) Technological development and new infrastructures for the future

As technological developments for improving comfort, speed, energy efficiency and the environment progress, financial systems must be adopted to support them. For developing such techniques as magnetic levitated linear motor trains, electric automobiles, low floor buses and trams, etc., ways of underwriting risk and securing social benefit should be devised. These projects should not be discarded because of defects in the financial systems.

(5) Roads as infrastructure and the social cost of automobiles

Roadside markets and large public gatherings are two characteristic uses of the road as a social space. Roads have also served as stages for street performers and as a playing space for children. And while these uses are still relatively recent, the last few decades have seen a shift in the use of roads to predominantly automobile transportation.

While comfortable modern life depends on the convenience of automobiles, there are also large social costs incurred in their use and most cities are facing the limits of motorization. This paradox forces our generation to research new directions for transportation. The road pricing schemes in Oslo and Singapore are important models for the future. Traffic demand management activities for businesses and for individuals are also important.

However, road construction as a social infrastructure should not entirely be neglected. It is gasoline and diesel automobiles that cause the environmental cost, not the roads themselves. In spite of the necessity of reducing automobile traffic, it is still necessary to construct high quality roads for new transportation systems that will appear in the future. "The roads for automobiles" should be changed to "the roads for the future of mankind."

Epilogue

The people who visit a city must come into contact with the transportation of that city. Transportation can be called "the face of the city." Just as many people imagine the cable car system when they think of San Francisco, transportation can form one of the major attractions of a city. A city with a beautiful, convenient and sometimes fun transportation system can create a good impression on visitors. Moreover, in the future, transportation will be a mirror reflecting a population's attitude toward environmental, social welfare, and safety issues, etc. The basic stance and daily activities of the citizens in regard to such matters will be known by observing their transportation system. The quality of transportation will be an important index measuring advancement in such areas.

For the past several decades, traffic demands and profitability have been prioritized in evaluating transport systems, and while such priorities may maximize the convenience and efficiency of a system, they cannot be expected to achieve the best results for societies at large, or for the global community. To construct an attractive transportation system as a face of the city is not easy. To construct a transportation system which is considered from the aspect of impact on the environment and friendliness to disabled individuals is not easy, either. However, if we view our transportation systems as an asset for improving the overall quality of life, we will see the need for careful evaluation of both the infrastructure itself, and the financial systems that support it.

SUMMARY OF INVESTMENT AMOUNTS AND COMPOSITIONS CALCULATED IN CHAPTER 5

Germany (Roadways)

Calculation objects

• Nearly all road improvements by the federal, state and local governments are covered.

Investment amounts

• Expenses for transport police are excluded.

• Expense for utility company (ÖFFA) are included.

Composition of Financial Resources

• Grants to local governments are not included in National Expenditures.

• Investment amounts excluding expenses by the federal government are appropriated for local expenditures.

• Bonds issued by ÖFFA are used as financial resources for road improvement but cannot be included as there is no actual mention of them in the statistical data.

France (Roadways)

Calculation objects

• Nearly all road improvements by the central and local governments, mixed-economy companies (SEMs) and so on are covered.

   Investment amounts ¥ Data coordination was attempted, and calculation made according to authorities such as the central government, local governments and mixed economy companies.

Composition of Financial Resources

• For local expenditures, the coordination of values mentioned in some data were attempted, and converted to 1993 values using values of National Expenditures.

• Improvement systems are taken into consideration and maintenance expenses were appropriated as Constructor's funds.

United Kingdom (Roadways)

Calculation objects

• Nearly all road improvements by the central government and local governments are covered.

Investment amounts

• Calculated using the data of International Road Federation.

Composition of Financial Resources

• Calculated using the data of International Road Federation.

United States (Roadways)

Calculation objects

• Nearly all road improvements by the federal, state and local governments and public corporations are covered.

Investment amounts

• Credits and interest payments are excluded.

• Investment amounts in toll roads are included. However, it is possible that toll road improvements by public corporations have not been included.

Composition of Financial Resources

• Investment amounts calculated in this study are proportionally divided at the rate of each financial resource to account for total revenue.

Japan (Roadways)

Calculation objects

• All road improvements by the central and local governments, four public corporations related to roads, designated city road public corporations and local road public corporations are covered.

Investment amounts

• General road improvements, toll road improvements by the four road-related corporations, local governments, designated city road public corporations and local road public corporations are calculated.

- For toll road improvements by the four road-related corporations, local governments, designated city road public corporations and local road public corporations, the contents of expenses mentioned were investigated, and those contents believed to be allotted to road improvement are appropriated.

Composition of Financial Resources

- For the four road-related corporations, income contents mentioned were investigated, the full amounts of grants from central and local governments are appropriated, and the amounts of the remainder proportionally divided into owners/operators' funds and debt and calculated.

- For designated city road public corporations, maintenance and repair expenses are calculated giving consideration that they are covered by toll income.

Germany (Railways)

Calculation objects

- German Federal railway (DB), East German National Railway (DR)
*Not included in calculation objects

  - Public railways of local governments

  - Private railways

Investment amounts

- There is no clear mention of maintenance and repair expenses in operating expenses, therefore they are calculated using the value for "Maintenance expense" in the operating expenses of German Federal Railway Company (DBAG) in the 1994 accounts for "Materials".

Composition of Financial Resources

- For local expenditures, they are not appropriated because there was no clear mention.

- For debt, the increasing amounts of debt accumulation are appropriated.

France (Railways)

Calculation objects

- French National Railway (SNCF), Paris transport corporation (RATP)
*Not included in calculation objects

- Public railway of local governments
- Private railways

Investment amounts

- For SNCF, "Network of the major trunk lines", "Network in Ile-de-France area (general plan, special plan)", "Other investment expenses" and "External expenses (Maintenance repair)" are appropriated.

- For RATP, there are no items to correspond to maintenance and repair expenses in operating expenses (statements of profit and loss). However, there is an item, "Operating improvement", in capital investment, and this value is utilized as maintenance and repairing expenses.

Composition of Financial Resources

- Out of local expenditures of SNCF, as for operating grants, they are not appropriated because there is no clear mention.

- For RATP, the values mentioned in "Planned investment use total" and "Investment out of plan use total" do not agree with values of "Planned investment financial resources total" or "Investment out of plan financial resources total" respectively. Therefore, every item was calculated, and a coordination attempt made.

United Kingdom (Railways)

Calculation objects

- British Railroads Board (BR), London Transport (LT)
*Not included in calculation objects

    - Public railway of local governments
    - Private railways

Investment amounts

- About maintenance and repair expenses in the operating expenses of BR, the values of "Terminals" and "Track, signaling and telecommunications" are appropriated.

- For the rolling stock expenditures of BR, they are calculated using the rate for rolling stock expenditures in the capital investment values.

- "Total renewal expenditures" for the maintenance and repair expenses in the operating expenses of LT are appropriated.

Composition of Financial Resources

• For the debt of BR, "Increase/(decrease) in borrowings in year" are appropriated.

• For National Expenditures of LT, the amount excluding operation subsidies from the subsidies for "Capital grants (National)" are appropriated.

United States (Railways)

Calculation objects

• Class I Railroads, Amtrak, public transportation (Transit)

Investment amounts

• For maintenance and repair expenses in the operating expenses of Class I Railroads, they are calculated using the values for 1994.

• For maintenance and repair expenses in the operating expenses of Amtrak, "Maintenance of way" and "Stations" are appropriated.

• For Transit, only the values for the transport mode (Automated Guideway, Cable Car, Commuter Rail, Heavy Rail, Light Rail, Monorail) believed to be investment amounts for railways are appropriated.

• For maintenance and repair expenses in the operating expenses of Transit, the values for "Non-Vehicle Maintenance" are appropriated.

Composition of Financial Resources

• For National Expenditures of Class I Railroads, they are calculated using the values for "Transfer from Government Authorities".

• For local expenditures and debt of Class I Railroads, "-" is appropriated because there was no mention of any figures for those categories.

• For National Expenditures of Amtrak, the full amounts of government grants are appropriated.

Japan (Railways)

Calculation objects

• All constructors (Private railway companies (161/168), Teito Rapid Transit Authority, Public subways, Japan Railway Construction Public Corporation and Honshu-Shikoku Bridge Authority)

• Former National Railway (1986)

Investment amounts

• Investment amounts of the private railway companies were calculated using the values mentioned in "Investigation report of capital investment trend by enterprises related to transportation".

• The rolling stock expenditures of Teito Rapid Transit Authority and public subways are calculate as 6% of the investment amount.

• For the Japan Railway Construction Public Corporation and Honshu-Shikoku Bridge Authority, the contents of mentioned expenses were investigated, and contents believed to be allotted to railway improvement appropriated.

• The rolling stock expenditures of the former National Railway were calculated using the values for 1976–80.

• For maintenance and repair expenses in the operating expenses of the former National Railway, "repair expenses (facilities and electricity)" are appropriated.

Composition of Financial Resources

• National Expenditures and local expenditures of the private railway companies were calculated using the values mentioned in "Subsidies comprehensive bibliography".

• For Teito Rapid Transit Authority and public subways, Japan Railway Construction Public Corporation and Honshu-Shikoku Bridge Authority, the income contents mentioned were investigated. Full amounts of subsidies from the central and local governments amounts are appropriated, and the remaining amount proportionally divided between owners/operators' funds and debt, and calculated.

• For the debt of the former National Railway, the total debt balance was calculated. The reimbursement amounts in the year concerned were added to the amount excluding the balance of previous year from the balance in the year concerned and calculated.

Germany (Airports)

Calculation objects

• New Munich airport
*Not included in calculation objects

  • Other international airports
  • Local airports

• Construction of the new Munich airport began in 1980 and the airport opened in 1992 but, because there were construction delays of four years, we adjusted the construction term to eight years, and the investment amounts in a single year are calculated.

- Maintenance and repair expenses are not included.

Composition of Financial Resources

- "Investment" and "Investment preparation" are appropriated for National Expenditures and local expenditures according to investment ratio.

- "Inner reserves" and "Leasing" are appropriated for owners/operators' funds.

France (Airports)

Calculation objects

- All airports including ones in overseas territories

Investment amounts

- Each capital investment amounts of "Grand Aeroports", "Aeroports Moyens", "Aeroports Intermediares", "Petits Aeroport" "Autres Aeroports", "Aeroports d'Outre-Mer" and "ADP" are appropriated.

- Maintenance and repair expenses are not appropriated.

Composition of Financial Resources

- Subsidies from the European Economic Communities for National Expenditures are included in addition to subsidies from the central government.

- "Prepayment" is appropriated to debt in addition to debt.

United Kingdom (Airports)

Calculation objects

- British Airport Authority Public Limited Company (BAAplc)
*Not included in calculation objects

- Local airports

- The values, excluding investment amounts in "Heathrow Express", from investment amounts of BAA are appropriated.

- Maintenance and repair expenses are not appropriated.

Composition of Financial Resources

- For the composition of financial resources, full amounts are appropriated for owners/operators' funds because there is no clear statistical data mentioned.

United States (Airports)

Calculation objects

• Object airports (3,285 airports of 5,598 known general airports) of NPIAS and investment amounts in airport construction reported by the Federal Aviation Administration (FAA).

• It is impossible to separate the items for maintenance expense and operating expense for the object airports of NPIAS. Therefore, the values for "Operation & Maintenance" are appropriated for maintenance and repair expense.

• "Operation" and "Facilities and Equipment" of aviation control are appropriated for airport improvement by the FAA.

Composition of Financial Resources

• For debt, the difference between total expense and total revenue of object airports in NPIAS was appropriated.

Japan (Airports)

Calculation objects

• All airports investment amounts

• Investment amounts by a 5-year plan excluding New Tokyo and Kansai international airports, but including New Tokyo International Airport investment amounts, Kansai International Airport investment amounts, Airport improvement special accounts maintenance operating expense, local airport maintenance expenses are appropriated.

• The content of mentioned expenses for the New Tokyo and Kansai international airports were investigated, and contents believed to be allotted to airport improvement appropriated.

Composition of Financial Resources

• For the financial resources for airport improvement in the 5-year plan, the figures were calculated using the financial resources contents for airport improvement specific accounts.

But, as for the local expenditures, the amount excluding expenses for airport improvement specific accounts from the investment amounts of the 5-year plan are appropriated.

SUMMARY OF CONTRIBUTORS' SHARES FOR INVESTMENT AMOUNTS AND
COMPOSITIONS CALCULATED IN CHAPTER 6

Germany (Roadways)

Shares of Contributor

• The petroleum tax burden to users, 23.8 penny/l is calculated for specific funds of the federation and 2.7 penny/l is calculated as specific funds of the local governments. Also, 80% of the automobile tax collected by the local governments is allotted to specific funds.

• 62.9 penny/l of the petroleum tax revenue and 20% of the automobile tax revenue burdened to roads users are calculated as general funds.

Burden amount

• The reimbursement of debt is not appropriated because there is no clear mention of it in the data.

France (Roadways)

Shares of Contributors

• Automobile users taxes that are burdened to road users are calculated as general funds.

Burden amounts

• The reimbursement of debt, "Servicing of loans" is appropriated from the expenses of the central government.

United Kingdom (Roadways)

Shares of Contributors

• Automobile users taxes that are burdened to road users are calculated as general funds of the central government.

• For local expenditures, as the finances of local governments depend on grants (RSG) from the central government in many areas, general funds of the local governments that road users are burdened with are calculated using the same rate charged to road users that are burdened with the accounts for general tax revenues in the central government.

Burden amount

• Because roads are not improved using debt, the results for contributors' shares are the same.

United States (Roadways)

Shares of Contributors

• The revenue allotted for road improvements out of the revenues of the Highway Trust Fund are calculated as road improvement specific funds.

Also, revenue (excluding amounts allotted to specific funds for public transportation) allotted to non-road improvement is calculated as general funds burdened to road users.

Burden amount

• For the reimbursement of debt, "Bond Retirement" is appropriated.

Japan (Roadways)

Shares of Contributors

• Gasoline tax, petroleum gas tax, automobile tonnage tax, gas oil delivery tax, automobile acquisition tax, petroleum gas transferred tax and automobile tonnage transferred tax are calculated as specific funds for road improvement.

• Automobile tax and light vehicle tax are calculated as general funds which road users are burdened with.

Burden amount

• For the reimbursement of debt, calculations were made by dividing the reimbursement amounts of each public corporation proportionally at the rate of investment amount to total expenses for the four road-related public corporations.

Germany (Railways)

Shares of Contributors

• The portion allotted to public urban passenger transportation from the petroleum tax revenue is calculated as the specific funds which road users are burdened with.

Burden amount

- For the reimbursement of debt, calculations are made by dividing the reimbursement amount of DB/DR proportionally at the rate of investment amounts to total expenses.

France (Railways)

Shares of Contributors

- For local expenditures of the SNCF, they are supposed to be completely covered by the "Versement de Transport," and the full amount is appropriated for "Indirect beneficiaries".

- Local expenditures of the RATP are supposed to be completely covered by the "Versement de Transport," and the full amount is appropriated for "Indirect beneficiaries".

Burden amount

- For the reimbursement of SNCF debt, the reimbursement amount was divided proportionally at the rate of investment amount to total expenses.

- For the reimbursement of the RAPT debt, the reimbursement amount was divided proportionally at the rate of investment amount to total expenses.

United Kingdom (Railways)

Shares of Contributors

- All grants to the BR and LT are covered by the general funds in central government. Therefore, there is not tax burden on users.

Burden amount

- For the reimbursement of BR debt, the reimbursement amount was divided proportionally at the rate of investment amount to total expenses.

- For the reimbursement of LT debt, debt as financial resources has not been utilized in recent years.

United States (Railways)

Shares of Contributors

- For the taxes burden on road users, revenues are appropriated for urban railways. Therefore, there is no tax burden on users for Class I Railroads or Amtrak.

- For Transit, the "Amount for Mass Transit" as outlay from the Highway Trust Fund as specific funds is appropriated in the federal, state and local governments.

- For Transit, "State and Local Taxes Dedicated at their Source for Transit Capital/Operating Funds Applied" are appropriated as the specific funds which Payers of Local taxes are burdened with.

Burden amount

- For the reimbursement of Class I Railroads, Amtrak and Transit debt, no appropriations are made.

Japan (Railways)

Shares of Contributors

- There is no tax burden to users.

Burden amount

- Because there is no data on the reimbursement of constructor debt, the values calculated in the study by Ishibashi are used.

Germany (Airports)

Shares of Contributors

- There is no data on airport improvement specific funds. Therefore, they are calculated under the assumption that all public funds are covered by general funds collected regardless of transport facility use.

Burden amount

- For the reimbursement of debt, nothing is appropriated for reimbursement after opening the airport.

France (Airports)

Shares of Contributors

- Calculated assuming that "Grants from central government" is the full tax amount users are burdened with.

Burden amount

- The reimbursement of debt for capital investment is appropriated for the amount of debt reimbursement.

United Kingdom (Airports)

Shares of Contributors

- For BAAplc, grants from the central and local governments are not used as financial resources so it is not necessary to consider a tax burden for users.

Burden amount

- For the reimbursement of debt, nothing is mentioned, so no appropriations are made.

United States (Airports)

Shares of Contributors

- Subsidies from the federal government are assumed to be covered by the Airport and Airway Trust Fund (AATF), and this amount is calculated as airport improvement specific funds.

Burden amount

- For the reimbursement of debt, nothing is mentioned, so no appropriations are made.

Japan (Airports)

Shares of Contributors

• Aircraft fuel tax and aircraft fuel transferred tax are calculated as airport improvement specific funds.

Burden amount

• Because there is no data mentioning reimbursement of debt, the values calculated in the study by Ishibashi are used.

Appendix 1. Statistical data list.

Transport Facility	Country	Data Name	FY	Issue
Roadways	Germany	World Road Statistics	1994	International Roads Federation
		Verkehr in Zahlen	1992,1995	Der Bundesminister für Verkehr
		Data of Association of Western Germany Roads	1980	Association of Western German Roads
		Straßenbaubericht	1991,1993	Der Bundesminister für Verkehr
		Statistisches Jahrbuch	1994	Statistishes Bundesamt
	France	World Road Statistics	1994	International Roads Federation
		Mémento de la Route	1993	Ministére de l'Equipment, des Transports et du Tourisme Direction des Route
		Construction Situation of Inter-City Expressways in Every Country	1992	—
		Annuaire Statistique de la France	1994	Ministére de l'Economie
	United Kingdom	World Road Statistics	1994	International Roads Federation
		Transport Statistics Great Britain	1994	Department of Transport
		Transport Statistics Report Road Traffic Statistics Great Britain	1995	Department of Transport
		Annual Abstract of Statistics	1995	HMSO
	United States	World Road Statistics	1994	International Roads Federation
		Highway Statistics	1992,1993	Federal Highway Administration
		Statistical Abstract of the United State	1993	Department of Commerce, Bureau of Census
	Japan	Road Statistics Yearbook	1992~1994	Whole country road users conference
		Japan Highway Public Corporation Yearbook	1991~1994	Japan Highway Public Corporation
		Data of Japan Highway Public Corporation	—	—
		Metropolitan Expressway Public Corporation Yearbook	1992~1994	Metropolitan Expressway Public Corporation
		Hanshin Expressway Public Corporation Yearbook	1992~1994	Hanshin Expressway Public Corporation
		Honshuu-Shikoku Bridge Authority Data	—	—
		Reference Calculation Data for Local Taxes	1994	Ministry of Home Affairs
		Financial Statistics	1993	Ministry of Finance
Railways	Germany	Data of Foreign Railway Investigation - No.36	1993	JR Group
		Present Situation of Foreign Railways	1994	JR Group
		Verkehr in Zahlen	1992,1995	Der Bundesminister für Verkehr
		World Railway Investment	1991	IRJ
		Company Report	1994	Deutsche Bahn
		Bericht über das Geschäftsjahr	1993	Deutsche Bahn
	France	Data of Foreign Railway Investigation - No.38	1993	JR Group
		Present Situation of Foreign Railways	1994	JR Group
		Present Situation and Future of French National Railway	1992	—
		World Railway Investment	1991	IRJ
		Mémento de statistiques Exercice	1994	SNCF
		Overseas Urban Railway Transport Information - No.31	1995	Teito Rapid Transit Authority

Appendix 1. Statistical data list.

Transport Facility	Country	Data Name	FY	Issue
Railways	United Kingdom	Data of Foreign Railway Investigation - No.40	1993	JR Group
		Present Situation of Foreign Railways	1994	JR Group
		World Railway Investment	1991	IRJ
		Transport Statistics Great Britain	1994	Department of Transport
		London Transport Annual Report	1993	London Transport
		Overseas Urban Railway Transport Information - No.30	1995	Teito Rapid Transit Authority
	United States	Data of Foreign Railway Investigation - No.39	1993	JR Group
		Present Situation of Foreign Railways	1994	JR Group
		Statistical Abstract of the United State	1993	Department of Commerce, Bureau of Census
		World Railway Investment	1991	IRJ
		Railroads Facts	1994	Association of American Railroads
		Analysis of Class I Railroads	1994	Association of American Railroads
		Annual Report	1993	National Railroad Passenger Corporation
		Data Tables for the 1993 National Transit Database section 15 Report Year	1993	Federal Transit Administration
	Japan	World Railway Investment	1991	IRJ
		Investigation Report of Capital Investment Trend by Enterprises Related to Transportation	1992,1993	Ministry of Transport
		Subway	1991,1992	Association of Japan Subway
		Financial Tables	1990~1992	Japan Railway Construction Public Corporation
		Honshu-Shikoku Bridge Authority Data	—	—
		Comprehensive Bibliography of Subsidies	1990~1992	
		Japan National Railway Inspection Report	1976~1980 1982~1985	Japan National Railway
		Transport Almanac	1988	Transport Newspaper
Airports	Germany	Construction Technique and Operation of Major Airport Overseas	1993	JAPIC
		Verkehr in Zahlen	1992,1995	Der Bundesminister für Verkehr
	France	Activité des aéroports français Année	1993	Ministére de l'Equipment, des Transports et du Tourisme Direction Générale de l'Aviation Civile-Service des Bases Aériennes Bureau de la Gestion des Aéroports
	United Kingdom	Transport Statistics Great Britain	1994	Department of Transport
		Construction Technique and Operation of Major Airport Overseas	1993	JAPIC
	United States	Report to Congress National Plan of Integrated Airport Systems (NPIAS) 1993-1997	—	Federal Aviation Administration
		FAA Statistical Handbook of Aviation	1993	Federal Aviation Administration

Appendix 1. Statistical data list.

Transport Facility	Country	Data Name	FY	Issue
Airports	Japan	Seeing Aviation by the Figures	1991,1994	Aerial Promotion Fund
		Data of New Tokyo International Airport	—	—
		Airport Handbook	'92~'93	Kansai Airport Investigation Meeting
		Financial Statistics	1993	Ministry of Finance
		Annual Statistics Report on Local Government Finance	1992~1994	Local Government Financial Society

Appendix 2.1(1). Investment amounts in road improvements (Table 5.1(1) contents in U.S.$).

		Investment amounts	National expenditures	Local expenditures	Owners / Operators' funds	Debt
Germany	Total	21,491	9,859	25,090		
	Bundes autobahnen (Autobahn)	3,249	2,953	296		
	Bundesund LandesStraßen (Federal and state roads)	5,660	2,711	2,949		
	Kreis straßen (District roads)	946				
	Gemeinde straßen (City, town and village roads)	9,321	398	12,184		
	Verwaltung und Sonstiges (Administration, others)	2,315				
France	Total	12,177	1,885	7,787	431	2,073
	Central and local governments	9,672	1,885	7,787		
	Toll Roads	2,505			431	2,073
United Kingdom	Total	9,418	3,846	5,573	0	0
United States	Total	75,338	15,523	48,737	2,998	8,080
	On State-Administered Highways	35,890	7,395	23,218	1,428	3,849
	On Locally Administered Roads	24,348	5,017	15,751	969	2,611
	Others	15,100	3,111	9,768	601	1,619
Japan	Total	101,826	25,504	54,436	7,441	14,445
	General roads and so on	78,303	24,448	53,855	0	0
	Local road public corporations and so on	1,792	0	388	40	1,365
	4 public corporations	21,731	1,056	193	7,402	13,080

Exchange Rate			
	Germany	1992	$0.615
	France	1993	$0.171
	U.K.	1991	$1.964
	U.S.A.	1992	$1.000
	Japan	1992	$0.008

Appendix 2.1(2). Investment amounts in road improvements (financial resource composition, %).

	Investment amounts	National expenditures	Local expenditures	Owners / Operators' funds	Debt
Germany	100.00%	28.21%	71.79%	0.00%	0.00%
France	100.00%	15.48%	63.95%	3.54%	17.03%
United Kingdom	100.00%	40.83%	59.17%	0.00%	0.00%
United States	100.00%	20.60%	64.69%	3.98%	10.73%
Japan	100.00%	25.05%	53.46%	7.31%	14.19%

Appendix 2.2(1). Investment amounts in railway improvements (Table 5.2(1) contents in U.S.$).

		Investment amounts	National expenditures	Local expenditures	Owners / Operators' funds	Debt
Germany	Total	7,997	5,869	0	1,419	708
	Deutsche Bahn (DB)	4,117	2,050	0	1,378	690
	Deutsche Reichsbahn (DR)	3,879	3,819	0	42	19
France	Total	3,504	1,975	187	767	576
	SNCF	2,871	1,856	94	594	327
	RATP	633	118	92	173	249
United Kingdom	Total	3,514	1,462	0	1,809	243
	British Rail	2,574	892	0	1,439	243
	London Transport	940	569	0	370	0
United States	Total	9,651	1,771	1,430	6,451	0
	Class I Railroads	5,738	5	0	5,733	0
	Amtrak	512	480	0	31	0
	Transit	3,402	1,286	1,430	686	0
Japan	Total	13,711	1,551	1,323	5,474	5,363
	Private railways	7,774	223	92	4,446	3,013
	Public subways	3,743	398	1,028	636	1,682
	Japan Railway Construction Corporation	2,194	930	204	392	668
	Honshu-Shikoku Bridge Authority	0	0	0	0	0

Exchange Rate	Germany	1993	$0.575
	France	1993	$0.171
	U.K.	1993	$1.499
	U.S.A.	1993	$1.000
	Japan	1992	$0.008

Appendix 2.2(2). Investment amounts in railway improvements (financial resource composition, %).

	Investment amounts	National expenditures	Local expenditures	Owners / Operators' funds	Debt
Germany	100.00%	73.39%	0.00%	17.75%	8.86%
France	100.00%	56.36%	5.32%	21.88%	16.44%
United Kingdom	100.00%	41.60%	0.00%	51.48%	6.92%
United States	100.00%	18.35%	14.81%	66.84%	0.00%
Japan	100.00%	11.31%	9.65%	39.93%	39.11%

Appendix 2.3(1). Investment amounts in airport improvements (Table 5.3(1) contents in U.S.$).

		Investment amounts	National expenditures	Local expenditures	Owners / Operators' funds	Debt
Germany	Munich	653	16	46	246	346
France	Total	476	64	29	208	175
	Aéroport de Paris (ADP)	295	4	0	163	128
	Others	182	60	29	45	47
United Kingdom	BAAplc	430	0	0	430	0
Unites States	Total	17,192	8,654	0	7,775	763
	Federal Aviation Administration	6,754	6,754	0	0	0
	Others	10,438	1,900	0	7,775	763
Japan	Total	6,078	1,104	228	1,850	2,897
	New Tokyo International Airport	703	180	0	89	434
	Kansai International Airport (KIX)	1,851	410	103	72	1,266
	Others	3,525	514	125	1,688	1,197

Exchange Rate			
	Germany	1992	$0.615
	France	1993	$0.171
	U.K.	1991	$1.964
	U.S.A.	1992	$1.000
	Japan	1992	$0.008

Appendix 2.3(2). Investment amounts in airport improvements (financial resource composition, %).

	Investment amounts	National expenditures	Local expenditures	Owners / Operators' funds	Debt
Germany	100.00%	2.45%	6.96%	37.65%	52.94%
France	100.00%	13.48%	6.07%	43.64%	36.81%
United Kingdom	100.00%	0.00%	0.00%	100.00%	0.00%
United States	100.00%	50.34%	0.00%	45.22%	4.44%
Japan	100.00%	18.16%	3.75%	30.43%	47.66%

Germany : Values of Munich International Airport
U.K.　　: Values of BAAplc

Appendix 3.1(1). Shares of contributors for road improvements (Table 6.4(1) contents in U.S.$).

Country	Payers of National Taxes	Payers of Local Taxes	Users (General taxes)	Users(Specific taxes)		Users (Fees)	Indirect Beneficiaries	Users (Later years)	Investment amounts
				Same mode	Other modes				
Germany	0	7,957	64	13,471	0	0	0	0	21,491
France	1,709	6,292	1,672	0	0	431	0	2,073	12,177
United Kingdom	3,179	4,607	1,632	0	0	0	0	0	9,418
United States	1,981	20,719	339	41,221	0	2,998	0	8,080	75,338
Japan	4,155	35,934	1,727	38,124	0	7,441	0	14,445	101,826

Exchange Rate	Germany	1992	$0.615
	France	1993	$0.171
	U.K.	1991	$1.964
	U.S.A.	1992	$1.000
	Japan	1992	$0.008

Appendix 3.1(2). Shares of contributors for road improvements (financial resource contents, %).

Country	FY	Payers of National Taxes	Payers of Local Taxes	Users (General taxes)	Users(Specific taxes)		Users (Fees)	Indirect Beneficiaries	Users (Later years)	Investment amounts
					Same mode	Other modes				
Germany	1992	0.00%	37.02%	0.30%	62.68%	0.00%	0.00%	0.00%	0.00%	100.00%
France	1993	14.03%	51.67%	13.73%	0.00%	0.00%	3.54%	0.00%	17.03%	100.00%
United Kingdom	1991	33.75%	48.91%	17.33%	0.00%	0.00%	0.00%	0.00%	0.00%	100.00%
United States	1992	2.63%	27.50%	0.45%	54.71%	0.00%	3.98%	0.00%	10.73%	100.00%
Japan	1992	4.08%	35.29%	1.70%	37.44%	0.00%	7.31%	0.00%	14.19%	100.00%

Appendix 3.2(1). Shares of contributors for railway improvements (Table 6.5(1) contents in U.S.$).

Country	Payers of National Taxes	Payers of Local Taxes	Users (General taxes)	Users(Specific taxes)		Users (Fees)	Indirect Beneficiaries	Users (Later years)	Investment amounts
				Same mode	Other modes				
France	1,975	0	0	0	0	767	187	576	3,504
United Kingdom	1,462	0	0	0	0	1,809	0	243	3,514
United States	951	1,318	0	0	1,210	6,172	0	0	9,651
Japan	975	1,323	0	0	0	6,050	0	5,363	13,711
Germany (DB/DR)	5,347	0	0	0	522	1,419	0	708	7,997
France (SNCF)	1,856	0	0	0	0	594	94	327	2,871
United Kingdom (BR)	892	0	0	0	0	1,439	0	243	2,574
United States (Class I, Amtrak)	485	0	0	0	0	5,764	0	0	6,250
Japan (JNR)	267	0	0	0	0	3,222	0	2,174	5,663

Exchange Rate			
	Germany	1993	$0.575
	France	1993	$0.171
	U.K.	1993	$1.499
	U.S.A.	1993	$1.000
	Japan	1992	$0.008

Appendix 3.2(2). Shares of contributors for railway improvements (financial resource contents, %).

Country	FY	Payers of National Taxes	Payers of Local Taxes	Users (General taxes)	Users(Specific taxes)		Users (Fees)	Indirect Beneficiaries	Users (Later years)	Investment amounts
					Same mode	Other modes				
France	1993	56.36%	0.00%	0.00%	0.00%	0.00%	21.88%	5.32%	16.44%	100.00%
United Kingdom	1993	41.60%	0.00%	0.00%	0.00%	0.00%	51.48%	0.00%	6.92%	100.00%
United States	1993	9.86%	13.66%	0.00%	0.00%	12.54%	63.95%	0.00%	0.00%	100.00%
Japan	1992	7.11%	9.65%	0.00%	0.00%	0.00%	44.12%	0.00%	39.11%	100.00%
Germany (DB/DR)	1993	66.87%	0.00%	0.00%	0.00%	6.52%	17.75%	0.00%	8.86%	100.00%
France (SNCF)	1993	64.66%	0.00%	0.00%	0.00%	0.00%	20.67%	3.28%	11.39%	100.00%
United Kingdom (BR)	1993	34.66%	0.00%	0.00%	0.00%	0.00%	55.89%	0.00%	9.45%	100.00%
United States (Class I, Amtrak)	1993	7.76%	0.00%	0.00%	0.00%	0.00%	92.24%	0.00%	0.00%	100.00%
Japan (JNR)	1986	4.72%	0.00%	0.00%	0.00%	0.00%	56.89%	0.00%	38.39%	100.00%

Appendix 3.3(1). Shares of contributors for airport improvements (Table 6.6(1) contents in U.S.$).

Country	Payers of National Taxes	Payers of Local Taxes	Users (General taxes)	Users(Specific taxes)		Users (Fees)	Indirect Beneficiaries	Users (Later years)	Investment amounts
				Same mode	Other modes				
France	39	29	0	25	0	208	0	175	476
United States	2,250	0	0	6,404	0	7,775	0	763	17,192
Japan	524	123	0	685	0	1,777	72	2,897	6,078
Germany (Munich)	16	46	0	0	0	246	0	346	653
France (ADP)	0	0	0	4	0	163	0	128	295
United Kingdom (BAAplc)	0	0	0	0	0	430	0	0	430
Japan (NRT)	85	0	0	94	0	89	0	434	703
Japan (KIX)	195	55	0	263	0	0	72	1,266	1,851

* Japan (NRT) : Values of New Tokyo Internatilnal Airport, Japan (KIX) : Values of Kansai International Airport

Exchange Rate			
	Germany	1992	$0.615
	France	1993	$0.171
	U.K.	1991	$1.964
	U.S.A.	1992	$1.000
	Japan	1992	$0.008

Appendix 3.3(2). Shares of contributors for airport improvements (financial resource contents, %).

Country	FY	Payers of National Taxes	Payers of Local Taxes	Users (General taxes)	Users(Specific taxes)		Users (Fees)	Indirect Beneficiaries	Users (Later years)	Investment amounts
					Same mode	Other modes				
France	1993	8.24%	6.07%	0.00%	5.23%	0.00%	43.64%	0.00%	36.81%	100.00%
United States	1992	13.09%	0.00%	0.00%	37.25%	0.00%	45.22%	0.00%	4.44%	100.00%
Japan	1992	8.63%	2.02%	0.00%	11.26%	0.00%	29.24%	1.19%	47.66%	100.00%
Germany (Munich)	-	2.45%	6.96%	0.00%	0.00%	0.00%	37.65%	0.00%	52.94%	100.00%
France (ADP)	1993	0.00%	0.00%	0.00%	1.28%	0.00%	55.25%	0.00%	43.47%	100.00%
United Kingdom (BAAplc)	1991	0.00%	0.00%	0.00%	0.00%	0.00%	100.00%	0.00%	0.00%	100.00%
Japan (NRT)	1992	12.14%	0.00%	0.00%	13.41%	0.00%	12.71%	0.00%	61.74%	100.00%
Japan (KIX)	1992	10.53%	2.98%	0.00%	14.19%	0.00%	0.00%	3.91%	68.39%	100.00%

* Japan (NRT): Values of New Tokyo International Airport, Japan (KIX): Values of Kansai International Airport

Appendix 4(1). Items included in investment amounts in road improvements.

Items	Germany	France	U.K.	U.S.A.	Japan
Expressways New construction	A	A	A	A	A
Site	A	C	B	C	C
Reconstruction, improvement, renewal	A	A	A	B	A
National highways New construction	A	A	A	A	A
Site	A	C	B	C	C
Reconstruction, improvement, renewal	A	A	A	B	A
Local roads New construction	A	B	A	A	A
Site	B	C	B	C	C
Reconstruction, improvement, renewal	A	B	A	B	A
Administration	B	B	A	A	C
Investment, design	A	A	A	A	A
Safety measure	A	B	C	A	C
Maintenance and repair	A	A	A	A	A

A : Items which are mentioned in statistical data clearly and included.
B : Items which are not mentioned in statistical data clearly but to think that these are included.
C : Items which can not be specifically grasped not to be mentioned in statistical data.
D : Items which are not included.

Appendix 4(2). Items included in investment amounts in railway improvements.

Items	Germany	France	U.K.	U.S.A.	Japan
Way and structures	A	A	B	A	A
New route construction	B	A	B	B	A
Stations	B	AB[1]	AB[2]	BAB[4]	B
Machines	A	A	C	C	C
Research and development	C	C	C	C	AC[5]
Maintenance and repair	A	A	BA[3]	A	A

A : Items which are mentioned in statistical data clearly and included.
B : Items which are not mentioned in statistical data clearly but to think that these are included.
C : Items which can not be specifically grasped not to be mentioned in statistical data.
D : Items which are not included.

Note 1)   SNCF ..... B,   RATP ..... A
Note 2)   BR ..... A,   LT ..... B
Note 3)   BR ..... B,   LT ..... A
Note 4)   Class I Railroads ..... B,   Amtrak ..... A,   Transit ..... B
Note 5)   All constructors ..... A,   JNR C

Appendix 4(3). Items included in investment amounts in airport improvements.

Items	Germany	France	U.K.	U.S.A.	Japan
Capital investment	A	A	A	A	A
Aviation aiding facilities	B	B	B	A	A
Maintenance and repair	D	D	D	A	AD[1]

A : Items which are mentioned in statistical data clearly and included.
B : Items which are not mentioned in statistical data clearly but to think that these are included.
C : Items which can not be specifically grasped not to be mentioned in statistical data.
D : Items which are not included.

Note 1) Kansai International Airport ..... D

Appendix 5(1). Shares of contributors for road improvements (burden ratio).

Country	FY	Payers of National Taxes	Payers of Local Taxes	Users (General taxes)	Users(Specific taxes) Same mode	Users(Specific taxes) Other modes	Users (Fees)	Indirect Beneficiaries	Users (Reimbursement)	Burden amounts
Germany	1992	0	12,939	104	21,906	0	0	0	0	34,949
France	1993	10,007	36,845	9,789	0	0	2,526	0	7,100	66,267
United Kingdom	1991	1,618	2,345	831	0	0	0	0	0	4,795
United States	1992	1,981	20,719	339	41,221	0	2,998	0	4,589	71,847
Japan	1992	523	4,522	217	4,798	0	937	0	753	11,750

* Germany: Million DM, France: Million FF, U.K.: Million £, U.S.A.: Million US$, Japan: Billion Yen

Appendix 5(2). Shares of contributors for railway improvements (burden ratio).

Country	FY	Payers of National Taxes	Payers of Local Taxes	Users (General taxes)	Users(Specific taxes) Same mode	Users(Specific taxes) Other modes	Users (Fees)	Indirect Beneficiaries	Users (Reimbursement)	Burden amounts
France	1993	11,565	0	0	0	0	4,489	1,092	3,856	21,003
United Kingdom	1993	975	0	0	0	0	1,207	0	62	2,244
United States	1993	951	1,318	0	0	1,210	6,172	0	0	9,651
Japan	1992	123	167	0	0	0	761	0	1,260	2,310
Germany (DB/DR)	1993	9,297	0	0	0	907	2,468	0	2,407	15,079
France (SNCF)	1993	10,871	0	0	0	0	3,476	551	2,703	17,601
United Kingdom (BR)	1993	595	0	0	0	0	960	0	62	1,617
United States. (Class I, Amtrak)	1993	485	0	0	0	0	5,764	0	0	6,250
Japan (JNR)	1986	34	0	0	0	0	405	0	122	561

* Germany: Million DM, France: Million FF, U.K.: Million £, U.S.A.: Million US$, Japan: Billion Yen

Appendix 5(3). Shares of contributors for airport improvements (burden ratio).

Country	FY	Payers of National Taxes	Payers of Local Taxes	Users (General taxes)	Users(Specific taxes)		Users (Fees)	Indirect Beneficiaries	Users (Reimbursement)	Burden amounts
					Same mode	Other modes				
France	1993	230	169	0	146	0	1,217	0	474	2,237
United States	1992	2,250	0	0	6,404	0	7,775	0	0	16,429
Japan	1992	66	15	0	86	0	224	9	143	543
Germany (Munich)	-	26	74	0	0	0	400	0	0	500
France (ADP)	1993	0	0	0	22	0	953	0	152	1,127
United Kingdom (BAAplc)	1991	0	0	0	0	0	219	0	0	219
Japan (NRT)	1992	11	0	0	12	0	11	0	28	61
Japan (KIX)	1992	25	7	0	33	0	0	9	0	74

* Germany: Million DM, France: Million FF, U.K.: Million £, U.S.A.: Million US$, Japan: Billion Yen
* Japan (NRT): Values of New Tokyo International Airport, Japan (KIX): Values of Kansai International Airport

BIBLIOGRAPHY

Abbott, K. and D. Thompson 1991, De-regulating European Aviation: The Impact of Bilateral Liberalisation, International Journal of Industrial Organization, Vol.9

Aberle, Gerd und Hedderich, Alexander 1993, Umsetzung der Bhanstrukturreform—Diskriminierungsfreier Netzzungang bei den Eisenbahnen, Internationales Verkehrswesen 45 Januar/Februar

Aeroports de Paris, Rapport d'activite

Akio Ono, Role and Functions of Railway Development Fund 1997, Japan Railway & Transport Review No.11

Aoki, Mami Transport and Fares Communities in the West Germany—Cooperation and Adjustment among Transport Companies, Annual Report on Transportation Economics, 1986

Aoki, Mami Provision of Infrastructure for Regional Transport and Regulation of Public Transport Services, Annual Report on Transportation Economics, 1991

Arakawa, Makoto, The Most Recent Status of Expressways in France, Expressways and Automobiles, Sep. 1991

Arnord, R. Douglas 1979, Congress and the Bureaucracy—A Theory of Influence, New Haven: Yale University Press

ASFA 1991, Bulletin des Autoroutes Francaisesdossier d'Information des Societes Concessionaires, No.32 AVRIL

Ashford, Norman, Airport Finance

Association of American Railroads, Railroads Fact

Association of American Railroads, Analysis of Class I Railroads

Association of American Railroads, Railroad Ten-Year Trends

B.S. Hoyle and R.D. Knowles, Modern Transport Geography, London

BAA, Report and Accounts

Bazex, Michel Contrats de Plan entre l'Etat et Enterprises Publiques, 1984

Bertrand Deveaud 1993, Le Grand Foutoir. Méfaits et aberrations qui paralysent l'Etat français

Blind, Wilhelm Heavy Spending on HS Infrastructure, International Railway Journal, 1990.6

Bonavia, Michael R, Twilight of British Rail?, David & Charles, 1985

Bonner Bericht 1991, Regierungskommission Bundesbahn legt ersten Zwischenbericht vor, Die Bundesbahn

Bonner Bericht 1990, Bei der Deutschen Reichsbahn gelten ab 3. Oktober die Bundesbahngesetze, Die Bundesbahn

Bonner Bericht 1992, Verkehrspolitische Kabinettsbeschlüsse zu Verkehrswegebau und Bahnreform, Die Deutsche Bahn

Bonner Bericht 1993, Bundestag beschleunigt Beratung der Bahnstructurreform-Gasetze, Die Deutsche Bahn

Brandborn, J. und Hellsvik, L. 1990, Die neue Eisenbahnpolitik in Schweden, Internationales Verkehrswesen 42. 6. Heft November/Dezember

British Railways Board, Annual Report and Accounts

British Road Federation, Basic Road Statistics

Bundesministerium für Verkehr 1985, Autobahnen in Deutschland

Bundesminister für Verkehr 1985, Bundesverkehrwegeplan 1985

Bundesminister für Verkehr 1990, Bundesverkehrwegeplan 1985 und Gesamtdeutscher Verkehrs-wegeplan

Bundesminister für Verkehr 1992, Bundesverkehrswegeplan

Bundesministerium für Verkehr 1993, Funfjahresplan für den ausbau der bundesfernstrassen in den jahren 1993 bis 1997 mit erganzung bis 2

Bundesminister für Verkehr 1990, Verkehrspolitik der 90er Jahre

Bundesminister für Verkehr 1991, Verkehrsprojekte Deutsche Einheit

Bundesminister für Verkehr 1992, Verkehrsprojekte Deutsche Einheit

Bundesministerium für Verkehr 1993, Verkehrsprojekte Deutsche Einheit

Bundesminister für Verkehr 1994, Verlehrspolitik

Bundesministerium für Verkehr, Verkehr in Zahlen

Bundesministerium für Verkehr, Starassenbaubericht

Bureau of Transportation Statistics, National Transportation Statistics

Bureau of Transportation Statistics, Transportation Statistics Annual Report

Bushell, Charis (ed.) 1995–1996, Jane's URBAN TRANSPORT SYSTEMS, Fourteenth Edition

Button and Swann 1991, Aviation Policy in Europe, in K. Button ed. Airline Deregulation: International Experience, Devid Fulton, London

Club de Bruxelles 1991, Transport in Europe: The Future of Inland Transport

Cole, Stuart Applied Transport Economics, Kogan Page, 1987

Collender Stanley 1986, The Guide to the Federal Budget, Washington, D.C.

Commission of the European Communities, Report of the High Level Group on the Development of a European High-Speed Train Network

Commission of the European Communities 1990, Communication on a Community Railway Policy, COM(89) 564 final

Confederation Française pour l'Habitation 1990, l'Urbanisme et l'Amenagement du Territoire, La France et ses Autoroutes

Congressional Budget Office 1988, New Directions for the Nation's Public Works

Congressional Budget Office 1985, US Congress, Toll Financing of US Highways, Dec.

Coste, Jean Francois Summary of Road Maintenance in France, Transportation Research Record 1183

Cubertafond, Alain 1993, Le Pouvoir, la politique et l'Etat en France,

degli Abbati, Carlo 1987, Transport and European Integration

Department of Transport 1983, Railway Finances: Report of a Committee chaired by Sir David Serpell, HMSO

Department of Transport, Transport Report

Department of Transport, Transport Statistics Great Britain, HMSO: Her Majesty's Stationery Office

Department of Transport, Transport Statistics Report Road Traffic Statistics Great Britain

Dernbach, Lothar 1985, Die Strategie DB '90- Entwicklung und Stand der Realisierung, Jahrbuch des Eisenbahnwesens

Deutsche Bahn 1994, The Railway Reform,

Deutsche Bahn, Facts and Figures

Deutsche Bahn, Moving With The Times

Deutsche Bahn, Company Report

Deutsche Bahn, Annual Report and Accounts

Deutsche Bahn, Bericht uber das Geschaftsjahr

Deutsche Bundesbahn 1991, Geschäftsbericht,

Doganis, R. 1989, Relagulatory Changes in International Air Transport, in K. Button and Swann, eds, The Age of Regulatory Reform, Oxford, U.K., Clarendon Press

Dreyfus G. 1982, Rapport sur la Situation actuelle et le Devenir des Autoroutes Françaises, Fevrier

Drocourt, M. L'Organisation de l'Exploitation des Autoroutes, No.660, 1990.12

Domergue, Philippe and Emile Quinet, Financing French High-Speed Network, Japan Ellwanger, G. 1990, Trennung von Netz und Betrieb bei den Ellwanger, G. 1990, Trennung von Netz und Betrieb bei den Eisenbahn, Internationales Verkehrswesen 42. 1. Heft Januar/Februar,

Ellwanger, G. und Hamelbeck, C., Das Schienennetz der DB-Bestandteil der statlichen Infrastruktur, Die Bundesbahn, 1989

Encaoua, D., Liberalizing European Airline: Cost and Factor Productivity Evidence, International Journal of Industrial Organization, Vol.9, 1991

European Communities, The European Highspeed Train Network, 1990.12

Federal Aviation Administration, FAA Statistical Handbook of Aviation,

Federal Aviation Administration, National Plan of Integrated Airport Systems (NPIAS) 1993–97,

Federal Aviation Administration, Annual Report,

Federal Aviation Administration Research and Special Programs Administration, Airpott Activity Statistics of Certificated Route Air Carriers,

Federal Highway Administration, America's Challenge for Highway Transportation in the 21st Century, Nov. 1988

Federal Highway Administration, Highway Statistics

Federal Highway Administrator, Toll Facilities in The United States

Federal Minister of Transport, Civil Aviation Policy Concept 2000

Gemeinschaft der Europäischen Bahnen, Vorschlag für ein Europäisches Hochgeschwindigkeitsnetz, 1989

Gestin, G. et B. Seligmann, L avenir du Secteur Autoroutier, PCM, Fevier 1984

Gibb. R and Charlton. C, International Surface Passenger Transport: Prospects and Potential, 1992

Gifford, Jonathan L., The Innovation of the Interstate Highway System, Transport Research A, vol.18A, No.4, 1984

Gifford, Jonathan L., The Saga of American Infrastructure: Toward the 21st Century, The Wilson Quarterly 17, 1993

Gifford, Jonathan L., Horan, Thomas A. and White, Louise G., Dynamics of Polivy Change: Reflections of the 1991 Federal Transportation Legislation, TRB 73rd Annual Meeting Draft, Preprint No.940814, 1994

Goodwin, P.B. Subsidized Public Transport and Demand for Travel, Oxford Studies in Transport, 1983

Group Transport 2000, Transport in Fast Changing Europe, 1990.1

Haanappel, P.P.C., G. Petsikas, R. Rosales and J. Thaler eds., EEC Air Transport Policy and Regulation, and Their Implications for North America, Kluwer Law and Taxation Publishers, 1989

Highway Routes, Expressways and Automobiles, May 1992

Hirayama, Kazuo Current State and New Directions of Toll Road Systems in Europe and the United States—Part I, Expressways and Automobiles, Oct. 1992

Hirayama, Kazuo Current State and New Directions of Toll Road Systems in Europe and the United States—Part II, Expressways and Automobiles, Nov. 1992

Hirayama, Kazuo Outline of the Intermodal Surface Transportation Infrastructure Act of 1991, Expressways and Automobiles, Feb. 1992

Hirayama, Kazuo The Highlights of Intermodal Surface Transportation Efficiency Act of 1991, Expressways and Automobiles, May 1992

HMSO, Transport Act of 1968, 1978, 1980, 1983

HMSO, Local Government Act of 1974

HMSO, Policy for Roads, England, 1980

HMSO: Her Majesty's Stationery Office, Annual Abstract of Statistics

International Railway Journal, World Railway Investment

International Road Federation, World Road Statistics

Hori, Masamichi Changes in the Management Improvement Plan and Management Strategy in the Germany Federal Railways, Annual Report on Transportation Economics, 1990

Hori, Masamichi International Comparison Study on Separation of Infrastructure and Operation in the Railway System, Annual Report on Transportation Economics, 1993

Hosaki, Yasuo The New Five-Year Road Improvement Program in the U.S.A., Expressways and Automobiles, Sep. 1991

Jäger, Heibert Projekt 'Vereinbarungen mit den Bundesländern und Marktmanagement Nahverkehr, Die Budesbahn, 1987

Jockwood, S.C., Unconventional Highway Financing in the United States, Report Prepaper for JETRO, 1990

Kakumoto, Ryohei Transportation Investment and Japan's Experience, Japan Railway & Transport Review No.11, April 1997

Kasoper, D.M., Deregulation and Globalization, Ballinger, 1988

Katayama, Kiyoshi Basic Plan for Public Investment for the Next Decade, Expressways and Automobiles, Sep 1990

Kosaka, Akihiro Yuichiro Kawaguchi, Yasuo Tomita, Yoshitsugu Hayashi 1993, A Comparative Study on Improvement and Management of Urban Railway Systems, Proceeding of Infrastructure Planning No.16(1) 1993.12

Kumano, Yoshimasa National Transportation Policy, Financial Innovations and Technological Demonstrations in the U.S.A.: From the Lecture of R.B. Robertson, Director General, International Road Federation, Expressways and Automobiles, April 1991

Kurashimo, Katsuyuki Some Comments of Interim Report of Road Council on Toll Road System, Expressways and Automobiles, Oct 1992

Lecomt, CH. 1990, Les Concession Autoroutieres Structures Juridques et Financiers, No.660

Lenz, C.O., The Decisions of the European Court of Justice on the Applicability of the Rule of the Treaty of Rome to Air Transport, in P.P.C. Haamappel et al. eds., 1989

Link, Heike Financing Rail Project in Germany, Japan Railway & Transport Review No.11, April 1997

London Transport, Annual Report

Lowe, D. 1989, The Transport and Distribution Manager's Guide to 1992, Kogan Page

Lowe, D. 1989, The Transport Manager's and Operator's Handbook 1989, Kogan Page

Mackie, P., D. Simon, and A. Whiteing, The British Transport Industry and the European Community, Gower 1987

MELATT—Direction des Investissements Routiers, Les clefs de financement, 1986.3

MELATT, La Lettre de l'Equipement, du Logement, de l'Aménagement du territoire et des Transports, No.20, 1987

Mendes de Leon, P.M.J., Euro-Cabotage: A Lever for Liberalization of International Civil Aviation, in P.P.C. Haamappel et al. eds., 1989

Ministère de l'Equipement, Le projet de shema directeur national des liasons ferroviaires a grande vitesse, 1989.1

Ministère de l'Equipement Direction des Routes, Mémento de la Route

Ministère de l'Equipement, des Transports et du Tourisme, Le Nouveau Schéma Directeur Routier National, 1990.1

Ministère de l'Equipement, des Transports et du Tourisme, Activité des aéroports français

Ministère de l'écnomie et des finances, Annuaire Statistique de la France

Ministry of Public Works, Housing, Territorial Planning and Transportation, General Presentation of the French Highway System

Mizoguchi, Makoto Construction Program of National Expressways in the Period of the 11th Five-Year Road Improvement Program, Expressways and Automobiles, June 1993

Nash, C.A. 1982, Economics of Public Transport, Longman

Nash, C.A. 1985, Playing Subsidy to British Rail: How to Get Value for Money, Transport UK 1985: An Economic, Social and Policy Audit, Policy Journals

Nash, C.A. and J. Preston, Railway Privatisation—United Kingdom, ECMT Round Table on Transport Economics

National Railroad Passenger Corporation (Amtrak), Annual Report

Observatoire Economique et Statistique des Transport, Mémento de Statistiques des Transport

OECD, Energy Price and Taxes, 1991

OECD Scientific Expert Group, Toll Financing and Private Sector Involvement in Road Infrastructure Development, Road Research Study, 1987

Pällman, Wilhelm Zukunftsstrategien der Bundesbahn im öffentlichen Personennahverkehr, Jahresbuch des Eisenbahnwesens, 1989

Planning Division, Road Bureau, Ministry of Construction, The Outline of the Recommendations of Road Policy Council: Designation of New National Highway Routes, Expressways and Automobiles, May 1992

RATP, Rapport Annuel

Regierungskommission Bundesbahn 1991, Bericht der Regierungskommission Bundesbahn

Rush W.A., Toll Highway Financing, Transportation Research Board, Dec. 1984

Sasamori, Hideki The Outline of the 11th Five-Year Road Inprovement Program, Expressways and Automobiles, June 1993

Schor, E., The Politics of International Aviation, Macmilan, 1991

Secretary of Transportation, The Draft Bill of Surface Transportation Assistance Act of 1991, Feb. 1990

Seligman, B. et. Justin, Presentation general du System autoroutier Françaises, le rapport presenté a OECD seminar, mars 1985

Smith, W.S. and N.H. Wuestefeld, Current Trends in Toll Financing, A Paper Presented at 62nd Annual Meeting of TRB, Jan. 1983

SNCF, Mémento de Statistiques Exercice 1994,

SNCF, Rapport d'Activité,

Starkie, D, The Motorway Age—Road and Traffic Politics in Post-war Britain, Statistiches Bundesamt, Statistisches Jahrbuch

Sugiyama, Masahiro The Federal Transportation Route Plan, Annual Report on Transportation Economics, 1982

Takano, Shigeru Outline of Report from Council for Transport Policy—Part I: Basic Concept of Transport Policy in 1990's Viewing at 21st Century, Expressways and Automobiles, Sep. 1991

Takeda, Fumio Evaluation of Toll Road System from National Economy Viewpoint and Comparative Study of Institutional Framework and Performance of 4 Major Toll Financing Countries, Annual Report on Transportation Economics, 1987

Tanaka, Hidetoshi Historical Perspectives of the U.S. Highway System Planning, Expressways and Automobiles, Jan. 1995

Tazaki, Tadayuki Recommendations on Toll Road System by Road Council, Expressways and Automobiles, Aug. 1992

Tokuyama, Hideo Long-Term Road Improvement Plan, Expressways and Automobiles, June 1993

Transport Committee 2nd Report, Future of the Railways in the Light of the Government's White Paper Proposals

Transport Committee Fourth Report, Railway Finances

Transport Press Limited, Railway Directory & Year Book, Railway Gazette International, 1990

U.S. Department of Commerce Bureau of the Census, Statistical Abstract of the United States

U.S. Department of Transportation, National Transportation Strategic Study, Mar. 1990

U.S. Department of Transportation, Moving America ... New Directions, New Opportunities, Feb. 1990

U.S. Department of Transportation, A Summary International Surface Transportation Efficiency Act of 1991, 1991.12

U.S. Department of Transportation, The National Highway System; The Backbone of American's Intermodal Transportation Network, Washington, D.C., 1993

U.S. Department of Transportation, Proposed National Highway System, Washington, D.C., 1993

Walton, C. Michael and Mark A. Euritt 1990, Highway Finance and The Private Sector-Issue and Alternative, Transportation Research Vol.24A, No.4

Wan, X., H. Nakamura, The Effects and Financial Policies of Expressway Development in the Industrialized Countries, Proceeding of Infrastructure Planning No.13 1990.11

Westfeld N.H. and E.J. Regan, III, Wilbur Smith and Associates, Impact of Rate Increases on Toll Facilities, presented at the 1981 IBTTA Workshop

Wildarsky, Aaran, The Politics of the Budgetary Process, Brown Company, 1984

Winghart, J.A. President de Saprr Travaux, Les Autoroutes Francaises et l'Europe, No.660, 1990.12

Yoshikawa, Yutaka New Directions of Public Works: From Congressional Budget Office Report of 1988, Expressways and Automobiles, Aug 1991